TREAT 'EM ROUGH!

Col. Ira C. Welborn, Tank Corps director in the United States. Welborn, an infantry officer, earned the Medal of Honor during the Spanish-American War.

TREAT 'EM ROUGH!

The Birth of American Armor, 1917–20

Dale E. Wilson

Foreword by
Maj. Gen. G. S. Patton, USA (Ret.)

PRESIDIO

Published by Presidio Press
31 Pamaron Way, Novato, CA 94949

Distributed in Great Britain by
Greenhill Books
Park House, 1 Russell Gardens
London NW11 9NN

Library of Congress Cataloging-in-Publication Data

Wilson, Dale, 1951–
    Treat 'em rough!: the birth of American armor, 1917–20 / Dale E.
Wilson.
        p.   cm.
    Includes bibliographical references.
    ISBN 0-89141-354-5
    1. United States. Army—Armored troops—History. 2. World War,
1914–1918—Tank warfare. I. Title.
D608.W54   1989
940.4'12'73—dc20   89-38468
                                                                    CIP

All photographs courtesy of the U.S. Army Signal Corps, except the photograph of Colonel Welborn courtesy of the United States Military Academy special collections.

All maps except that of the 301st Tank Battalion operations on 23 October 1918 adapted from American Battle Monuments Commission, *American Armies and Battlefields in Europe*. Washington, D.C.: U.S. Government Printing Office, 1938.

Map of 301st Tank Battalion operations on 23 October 1918 adapted from Edmonds and Maxwell-Hyslop, *History of the Great War*. London: His Majesty's Stationery Office, 1947.

Sketch in chapter II courtesy National Archives.

Sketch in chapter IX courtesy U.S. Army Military History Institute, Carlisle Barracks, Pennsylvania.

Printed in the United States of America

Contents

Foreword

About one year ago, when Dale Wilson asked me to review his manuscript, I sensed its excellence, and predicted that it would eventually be published as a full-scale historical volume. Happily, I was correct! In this respect, I am especially delighted that the selected publisher is my old friend and West Point 1946 classmate, Bob Kane. No other publishing house known to me can match the quality of Presidio Press as a producer of military history books.

An army son, I was born in December, 1923. Living in Washington with my parents from 1929 until 1934, I was virtually raised on first-hand stories of tank operations from the 1914–1918 War.

I can still recall vividly the parade of narrators who came to call on my father, their former commander. Some of these visitors were Sereno Brett, Sasse, Joe Viner, Elgin Braine, Arthur Snyder, Harry Semmes, and many more great personalities whose names escape my memory. As a young and impressionable sub-teenager, I was permitted to sit in a corner after dinner and listen to conversations involving training at Bourg and Langres, the initial engagement at St. Mihiel, and finally, the Argonne Forest where, near the village of Cheppy, my father was seriously wounded and carried from the field by an Italian-American soldier, named Joe Angelo.

The echoes of those war time reminiscences remain clear in my memory: "What an S.O.B. you were, George, but we loved you!" "Do you remember the time when . . . ?" "I couldn't get the damn thing started."

"I called the repair folks and they couldn't either!" "Hell, we didn't eat for three days." And so forth, into the late evening hours. My mother, Beatrice, literally would have to order me to bed, reluctant as I was to leave that fascinating company.

Theirs was the story of the birth of the tank in the United States Army. It is a wonderful tale, and thanks to Dale Wilson, it now has been pulled together and recorded for all time.

The auspicious birth of the tank notwithstanding, its near death soon followed, and is skillfully laid out for us in Wilson's vitally important epilogue. From 1920 to the early 1930's, the tank and its associated operational concepts (which later evolved into armor) were steadily shoved aside by the traditionalists. Accordingly, many of those who had pioneered this vastly important research effort either left the army or became terribly discouraged and returned to their former branch of service. Our national military and political leaders could not yet see the potential demonstrated by this noisy and cumbersome combat vehicle, and literally allowed it to die on the vine.

In time, of course, the picture changed. When the Nazi threat became more evident, even to the most casual observer, the combined arms team which was later to become known as armor was conceived and born at Ft. Knox, Kentucky. However, the men described here, whose early efforts contributed significantly to victory in Europe in 1945, for the most part received little recognition following that victory.

Now, after nearly 70 years, Captain Wilson has given us the opportunity to recognize and salute those pioneer tankers who made armor history, by "Treating 'em Rough." They indeed broke trail for those who followed on.

G. S. Patton
Maj. Gen., USA (Ret.)

Preface

As a career Army officer, I have long been fascinated with the study of military history. After receiving my commission in the Army's Armor Branch in 1979, I became particularly interested in the study of tanks and mechanized warfare – especially the development of mechanized theory and doctrine.

Abundant literature is available on the conduct of armored operations in the Second World War, the Korean conflict, Vietnam, and the Arab-Israeli wars. Historians have also devoted considerable attention to the post–World War I development of mechanized theory, with special emphasis on the work of England's Basil Liddell-Hart and Maj. Gen. John F. C. Fuller and of German Gen. Heinz Guderian. The creation of tanks in 1914 and their employment by the British in the First World War have also been explored in depth.

Unfortunately, historians have virtually ignored the American experience with tanks in World War I. The most notable work on the subject was done by Martin Blumenson in the first volume of his collection of *The Patton Papers*. But Blumenson's work, excellent though it is, goes little farther than the experiences of George S. Patton, Jr., with the American Expeditionary Forces Tank Corps in France. Similarly, Dwight D. Eisenhower's biographers have provided glimpses of the activities of the Tank Corps in the United States insofar as they involved Eisenhower, who commanded the first stateside tank training center at Camp Colt, near Gettysburg, Pennsylvania, in 1918. Studies of American operations in France contain little (usually erroneous) or no information about the employment of tanks.

This dearth of information about America's first experience with the tank started me on a path that culminated in this work. I must confess that it was a fascinating journey. Along the way I became acquainted with a number of men whose names have long since faded into obscurity: men like Majs. Sereno E. Brett and Ralph I. Sasse, Lt. Col. Joseph W. Viner, Cols. Henry E. Mitchell and Daniel D. Pullen, Capt. Elgin Braine, Brig. Gen. Samuel D. Rockenbach, and a host of others who shared a vision of what the tank could accomplish and worked to make that vision a reality on the battlefield.

Although the equipment with which they operated was primitive, subject to frequent mechanical failure, depressingly slow, and frightfully uncomfortable to maneuver in, it made a significant impact on tactical thinking—especially in the minds of Patton and Eisenhower, who went on to achieve much greater fame commanding vast mechanized formations in the next war.

It is one of the perplexities of the era that the nation that provided the technology spawning the idea for the tank did not, in fact, first produce that vehicle. Nevertheless, some American military men in France and the United States recognized the potential of early British and French tanks and encouraged their employment by the United States Army.

What follows is the story of the American Tank Corps in World War I—from its creation to its dismemberment after passage of the National Defense Act of 1920. Particular attention is devoted to the development of equipment, organization and tactics, and a training program, all of which had to be accomplished from scratch in order to prepare the tanks and the men who would use them for combat. The meat of the story, however, is contained in the detailed accounts of the 1st/304th Tank Brigade's support of the American First Army in the St. Mihiel and Meuse-Argonne campaigns, and the 301st Heavy Tank Battalion's combat experiences with the British Fourth Army beginning in late September 1918.

Readers who are appalled by the convoluted workings of America's present defense research, development, and procurement system—which requires a decade or more to bring a new weapons system from the drawing board to production—will experience a sense of déjà vu as they read the account of Captain Braine's frustrating dealings with the Ordnance Department and civilian manufacturers. Despite the fact that the Renault Corporation provided detailed plans so the Americans could produce copies of the French light tank, not one was delivered to combat units before the Armistice. The effort to produce a joint American-British heavy tank was equally futile. Although the level of technology has increased exponentially in the intervening seventy years, the bureaucratic machinations occurring behind the scenes have been little altered.

It is hoped that this work will serve not only as a detailed account of a neglected part of America's military history but also as a case study for military and civilian leaders faced with the difficult task of preparing new weapons systems for battlefield employment in this era of increasingly rapid technological change.

It is impossible to thank adequately all of the people who have made this work possible. However, I wish to express my gratitude to several persons whose assistance proved invaluable. I am particularly indebted to Maj. Gen. George S. Patton (United States Army, Ret.), who kindly granted me access to his father's personal papers in the Library of Congress. I would also like to thank David Keough of the archives of the Army's Military History Institute at Carlisle Barracks, Pennsylvania. His patient guidance as I began my fumbling first steps in the research process was especially welcome. Likewise, I owe special thanks to the many people who assisted me at the National Archives as I waded through the more than 150 feet of Tank Corps files stored there. David Holt at Fort Knox's Patton Museum of Cavalry and Armor was also of great assistance, helping me sort through the contents of the Joseph W. Viner Collection held there.

The maps and illustrations accompanying this work are the product of many long hours of painstaking labor performed by Ed Krasnaborski, the cartographer in the United States Military Academy Department of History. I wish to extend my special thanks to him for his patience in dealing with me as production deadlines approached and I continued to press him with additional pleas for assistance—which he most cheerfully and graciously provided.

Professors Russell F. Weigley, Waldo Heinrichs, Herbert Bass, and Kenneth Kusmer, members of my doctoral committee at Temple University in Philadelphia, are superb mentors who introduced me to the historical profession and showed me how to approach this project with professional detachment. Special thanks are due Professor Weigley for the long hours he spent poring over the manuscript. His skilled editing and probing questions contributed significantly to the shaping of the final product. I would also like to thank Professor Edward M. Coffman of the University of Wisconsin and Maj. Daniel P. Bolger, formerly of the United States Military Academy Department of History, for taking the time to read and comment on the manuscript. Their insights proved most helpful. It should go without saying, however, that any mistakes are solely of my own making.

Finally, I reserve my deepest gratitude for my lovely wife, Aurelia, who patiently endured my long absences (frequently physical, more often mental) during the many months I spent so deeply absorbed in this project.

Introduction

The bloody stalemate that settled over the Western Front in late 1914 taxed the best minds of the general staffs of both the Entente and the Central Powers as they sought a means to restore mobility to the battlefield. Unfortunately, the power of the tactical defense (aided principally by the machine gun) had become so immense as to make direct infantry assault suicidal. Armies conducting offensive operations found themselves pouring troops into a meat grinder that churned out casualties by the hundreds of thousands. "Successful" offensive gains were measured in feet and meters—not miles or kilometers.

By October 1914, British Lt. Col. Ernest D. Swinton, serving at the time as a correspondent with the British Expeditionary Forces, had reached the conclusion that an armored machine capable of forcing its way through barbed-wire obstacles, climbing over trenches, and destroying or crushing machine guns was needed if the stalemate was to be broken. Swinton, inspired by a letter from a friend who described the American Holt caterpillar as "a Yankee tractor which could climb like the devil," drafted a proposal that he forwarded to the War Office on 20 October calling for the construction of heavily armored caterpillar tractors armed with artillery pieces and machine guns.[1]

Although the reaction of many leaders to Swinton's proposal was less than enthusiastic, it fired the imagination of at least one powerful Englishman: Winston Churchill, the first lord of the Admiralty. In January 1915,

Churchill, anxious to get his Royal Naval Air Service (RNAS) involved on the Continent, ordered Capt. Murray Sueter, director of the Admiralty Air Department, to put his staff to work designing a vehicle capable of crushing trench works.[2]

During the months that followed, a number of experimental wheeled and caterpillar-tracked armored vehicles were developed and tested by officers of the Admiralty Air Department before Sir William Tritton and Lt. Walter G. Wilson of the RNAS made a major design breakthrough. The Tritton-Wilson vehicle was the first tank to be configured in the now-familiar rhomboidal shape with the track encircling the body. It featured a pair of sponsons designed by Sir Tennyson D'Eyncourt in which two six-pounder guns were mounted. This vehicle was demonstrated publicly on 26 January 1916 and is considered to be the first true British tank.[3] It quickly earned the nickname "Mother," and all subsequent tanks of this type were called "Big Willies."[4] Because of the Royal Navy's involvement in tank development, a number of nautical terms, such as hull, port, bow, and hatch, were used to designate various tank parts.

The British went to great lengths to conceal the existence of their "landships" from the enemy. Everyone in any way involved with the project was sworn to secrecy, and personnel suspected of discussing the project were threatened with internment under the Defence of the Realm Act. Women known to have been informed of the project were "told that if the secret reached the enemy thousands of lives would be lost. . . . [Other personnel] who knew about the existence of the Landship Committee were informed that all the experiments had failed, and that the people concerned had lost their jobs." It was a report that was readily accepted.[5]

To further protect the secret, the Landship Committee decided to change the vehicle's name out of fear the very word *landship* might betray the secret. One author describes how the new name was chosen in the following (probably apocryphal) story:

> In the earlier stages of the vehicles' manufacture the machine resembled a cistern or reservoir, and it was decided to call it a "water-carrier." . . . [But,] the secretary of the "Water Carrier" Committee thought that the new title would be highly unsuitable, if not ludicrous [if only the committee's initials were used to identify it, a common government practice]! The name was therefore changed to "tank," and the committee was called the "Tank Supply" or "T.S." Committee.[6]

A more widely accepted (though less colorful) explanation for how the tank got its name is that the British, in an effort to deceive the enemy, when shipping early models to France for battlefield testing, listed them on ships' manifests as water tanks en route to Russia.[7]

The French experimented with tank designs during this same period. The only similarity between their vehicles and those of the British, however, was the combination of firepower under armor with the added power of caterpillar traction. The tactical theories of the two allies differed radically, and so too did the designs of the tanks they produced.[8]

The British became the first to employ tanks in combat, deploying forty-nine Mark I models on 15 September 1916 during the Battle of the Somme. Their effectiveness was hampered by the fact that they were not employed en masse but were instead scattered piecemeal on the battlefield. As might have been expected with such a primitive mechanical design, breakdowns were frequent. Nevertheless, results were encouraging.[9]

On 16 November, the British used two tanks to lead the attack at Beaumont-Hamel. One crossed the Germans' front-line trench and became stuck, while the other became mired in front of the trench. Despite this fiasco, the Germans were so shocked by the tanks' appearance on the battlefield that soldiers in both the front line and supporting trenches began waving white cloths to signal their surrender. The tank crews and supporting infantry were able to capture the entire garrison before the Germans could discover that the tanks were immobilized and all but at their mercy.[10]

Inspired by the manner in which the British employed their tanks offensively, the French scrapped their plan to use tanks as troop carriers and decided instead to employ them as accompanying artillery.[11] This decision was reflected in the design of the Schneider and St. Chaumond tanks. The Schneiders made their battlefield debut on 16 April 1917, when 132 were deployed at the Chemin des Dames. The St. Chaumond was first used on 5 May 1917, with sixteen joining an attack at Laffaux Mill.[12]

The French learned that accompanying artillery with tractor power did not really require the armor of a tank, so they designed a lightweight, highly mobile, turreted tank to serve in the infantry support role.[13] This tank, the Renault Char FT (for "faible [light] tonnage"), featured a two-man crew (significantly smaller than the six- to nine-man crews employed in the heavier British and French tanks) and mounted either a single 37mm gun or an 8mm machine gun in its turret. This vehicle became the backbone of the French Tank Corps, although it was not used in combat for the first time until 31 May 1918.[14]

All of these developments captured the attention of the chief of the U.S. Army War College, who had seen reports on tank developments submitted by the American Military Mission in Paris. While most of those reports had been highly critical of early tank operations, and the Paris-based observers declared tanks a failure (which became the official position of the War Department in early 1917),[15] the War College director ordered the Mission to report on the latest British and French tank theories and operations. That report, dated 21 May 1917, included the personal observations of Maj. Frank Parker, liaison officer at the headquarters of the French Armies of the North and North-East, on French tank operations in the April offensive.[16] This report would have significant influence on the future of tanks in the United States Army during World War I.

Notes

1. F. Mitchell, *Tank Warfare: The Story of Tanks in the Great War* (London: Thomas Nelson and Sons, Ltd., 1933), 4.

2. Rear Adm. Sir Murray Sueter, *The Evolution of the Tank* (London: Hutchinson and Company, Ltd., 1937), 53; ibid., 5–6.

3. Sueter, *Evolution of the Tank,* 89–90.

4. Mitchell, *Tank Warfare,* 9.

5. Ibid., 10.

6. Ibid., 11.

7. Martin Blumenson, *The Patton Papers, 1885–1940* (Boston: Houghton Mifflin Co., 1972), 436.

8. Capt. Joseph W. Viner, *Tactics and Techniques of Tanks* (Fort Leavenworth, Kansas: The General Service Schools Press, 1920), 4.

9. Maj. Ralph E. Jones, Capt. George H. Rarey, and 1st Lt. Robert J. Icks, *The Fighting Tanks since 1916* (Washington, D.C.: The National Service Publishing Co., 1933), 5–9.

10. Ibid., 9.

11. Viner, *Tactics and Techniques,* 5.

12. Jones et al., *The Fighting Tanks,* 55, 60.

13. Viner, *Tactics and Techniques,* 5.

14. Jones et al., *The Fighting Tanks,* 63–64.

15. *U.S. Army in the World War, 1917–1918: Organization of the AEF, Vol. 1* (Washington, D.C.: Historical Division, U.S. Army, 1948), 138.

16. Brig. Gen. Samuel D. Rockenbach, "Operations of the Tank Corps, A.E.F., with the 1st American Army" (Dec. 1918), U.S. Army Military History Institute, Carlisle Barracks, Pa. (hereafter, USAMHI), 1.

PART I

"If there was anything [Patton] wanted it was to
make the Tank Corps tougher than the Marines and
more spectacular than the Matterhorn."

—2d Lt. Will G. Robinson
311th Tank Center
Spring 1918

CHAPTER I

The Birth of the Tank Corps

The beginnings of the American Expeditionary Forces (AEF) Tank Corps can be traced to June 1917, when, shortly after arriving in Paris, Gen. John J. Pershing read a copy of the American Military Mission's 21 May report on British and French tank tactics and operations and was favorably impressed. Pershing, commander in chief of the AEF, immediately appointed several committees to study tank warfare, and some of his staff members were detailed to go to the front lines to study British and French equipment, organization, and tactics.

Initial reports from Pershing's staff indicated that early operations had been marred by numerous mechanical failures but that the effects of the tanks on the enemy more than compensated for their mechanical shortcomings. Despite the misgivings of some officers, Pershing thought that British-style heavy tanks and French light tanks could prove to be valuable assets to the AEF.[1]

All observers agreed that the French Schneider and St. Chaumond vehicles were unsatisfactory. Neither vehicle could truly be classified as a tank. Instead, they were nothing more than armored artillery carriers requiring infantry skirmishers to lead them into battle, carefully marking the routes they should follow. Underpowered and lightly armored, they did poorly traveling cross-country, and their crews suffered badly if they received direct hits from artillery fire. Another factor contributing to the decision to further investigate the British heavy tanks and French Renault tanks was

9

the inability of members of a joint French-British tank board to reconcile their theories on tactics and equipment when that body met in London in May 1917. The British insisted on using the heavy tank to clear the way for the infantry, while the French argued that light tanks operating in close liaison with the infantry offered the optimum battlefield solution. Their concept was to deploy the Renaults with the battalion support, advancing them only when the infantry assault bogged down.[2]

On 19 July Pershing ordered the creation of an American tank board to perform a detailed study of the French Renault and British heavy tanks. The board's members, Cols. Fox Conner and Frank Parker, Lt. Col. Clarence C. Williams, and Maj. Nelson E. Margetts, were decidedly protank, and their findings had significant influence on subsequent events.[3]

Ten days later, after being advised of Pershing's decision to have tanks in the AEF, the AEF's chief ordnance officer requested information on the number that would be required so a requisition could be passed on to the War Department in Washington. In response, Pershing ordered Lt. Col. LeRoy Eltinge, a member of his staff, to take charge of all tank matters and accomplish this task.[4]

The members of the tank board submitted a report containing their findings on 1 September. They concluded that the tank was "destined to become an important element in this war" and that a separate Tank Department operating under a single chief who would report directly to Pershing should be organized immediately. They further observed that of all the tank types then in production or being planned, only the French Renault and British Mark VI (a twenty-seven- to thirty-ton heavy tank that never reached production) could be expected to provide satisfactory results. Based on a projected strength of twenty combat divisions, the board's members recommended that a fleet of 2,000 light and 200 heavy tanks be procured, with production geared to provide for a 15 percent per month replacement rate.[5]

Armed with the board's findings, Eltinge set out to draft specific requirements for a "Combat Tank Service" for the AEF. Working in close coordination with other members of the AEF staff, Eltinge determined the number of tanks that would be required, the number and type of units, and the number of personnel needed to man the force. He based his recommendations on the needs of an army consisting of twenty fighting and ten replacement divisions.[6]

Pershing approved Eltinge's preliminary recommendations and directed him to immediately notify the War Department's chief of ordnance of the

AEF's tank requirements. Eltinge dispatched the following cable on 14 September:

> Careful study French and British experience with tanks completed and will be forwarded by early mail. Project includes 350 heavy tanks of British Mark VI pattern; 20 similar tanks equipped for signal purposes; 40 similar tanks for supply of gasoline and oil; 140 tanks arranged to carry 25 soldiers or five tons supplies; 50 similar tanks with upper platform for field gun; total 600 heavy tanks. Also following Renault tanks: 1,030 for fighting purposes; 130 for supply; 40 for signal purposes; total 1,200 Renault tanks. Replacement of tanks requires 15 per cent per month after arrival here.[7]

Eltinge further recommended that the Mark VIs be produced in two versions, one mounting a six-pounder gun and four machine guns, the other mounting six machine guns. It was suggested that the armament for the Renaults be either a single machine gun, six-pounder, or three-inch gun, with production to be fixed at a 2:1 ratio in favor of machine guns. A number of automotive requirements were also listed, including 300 six-ton trucks for transporting Renault light tanks, ninety three-ton trucks, 270 three-ton trucks with trailers, ninety three-ton trucks with kitchen trailers, ninety Ford automobiles, and 180 motorcycles.[8]

Eltinge also reported that the French were willing to permit manufacture of the Renault tank in the United States and that the Renault Works would supply a model to facilitate production. In exchange, the French desired 2,000 Renaults from the United States. The British, in the same spirit of cooperation, agreed to provide complete plans and specifications for their Mark VI tank so that its production could also begin in America.[9]

Organizational and personnel requirements were included in a detailed memorandum sent to the War Department on 23 September. This document requested authorization for thirty light tank companies for division troops; thirty light and fifteen heavy tank companies, five carrier companies, and two artillery carrier companies for army troops; ten training and replacement companies; five repair and salvage companies; a depot company; and support troops for army headquarters and general headquarters (GHQ). It was estimated that 14,827 soldiers would be needed to man these organizations.[10]

Little was accomplished during the next three weeks, but then, on 14 October, Majs. James A. Drain and Herbert W. Alden were detailed by the chief of ordnance in Washington to gather more information on the use, design, and production of tanks. [11]

Between 16 October and 4 November, Drain and Alden toured a number of French and British tank facilities, studying production, training, supply, and repair and salvage operations. At Circotte, the main French tank training camp and supply depot, Drain observed that the French were "laying this place out on a very large scale, evidencing an intention to make the tank a large and important arm of the service." [12]

On 2 November the two officers met in Paris with other Allied tank experts to discuss a common tank design program that would involve the United States, England, and possibly France. A tentative design for a heavy tank was agreed upon with the British, and the French exhibited interest in obtaining "at least a moderate number" of the proposed machines. [13]

Drain and Alden submitted a detailed report of their findings on 10 November. In the report they recommended that the United States accept the Renault tank as it was except for the turret, which they thought should be made of armor plate instead of cast iron. They also recommended that the United States produce only one type of heavy tank. Because Drain and Alden thought all of the British heavy tanks then in existence were inadequate, they proposed a joint British-American effort, with a detailed design to be "worked out at once in England" by engineers from the two countries. Production, they said, could best be accomplished in France, with the British supplying the armor and armament and the Americans providing Liberty twelve-cylinder aircraft engines and other automotive parts. [14]

Drain and Alden believed, however, that it would be useless to have good tanks without good men to operate them. They observed (somewhat jingoistically) that Americans would make good tank crewmen because they "are a strong race" and "of good character." [15] Drain and Alden recommended that personnel with a "high standard of fighting quality" be carefully selected from the ranks of these "good men" and that a comprehensive, "carefully worked out plan for training [them] be inaugurated at once." [16]

Finally, the two officers proposed that an American tank commissioner be appointed, "clothed with sufficient authority to enable him to fully represent our forces," and that an advisory staff of American, French, and British officers responsible for formulating a program for the combined use of tanks be created. [17] This latter recommendation was acted on almost immediately. Shortly after the report was filed, the Inter-Allied Tank Commission was organized, with Drain appointed to act as the American representative. He was ordered to seek "an agreement with the British and French as to the best type of tank to be constructed and coordinate the production effort so as to get the largest number of Tanks in the minimum time."[18]

The growing interest in tanks in the American ranks caught the attention of a young cavalry captain serving on General Pershing's staff. This officer, George S. Patton, Jr., was none too happy with his job as post adjutant and commander of the AEF Headquarters Company at Chaumont. Patton expressed this dissatisfaction in a letter to his wife in September. He explained that he was interested in all of the talk about tanks because he could "see no future to my present job." While he had heard that the tanks themselves suffered a high attrition rate, "the people in them are pretty safe." Not wanting to alarm her further, he observed that it would "be a long time yet before we have any [tanks]" and that they would have plenty of opportunity to discuss the matter before he could submit an application for tank duty.[19]

But the opportunity to work with tanks presented itself sooner than he had expected. In an undated diary entry, Patton recalled that he was approached "about the end of September" by Colonel Eltinge, who asked if Patton wanted to become a tank officer. "I said yes and also talked the matter over with Col. [Frank] McCoy [the assistant chief of staff] who advised me to write a letter asking that in the event of Tanks being organized that my name be considered."[20]

Patton followed up on this recommendation on 3 October with a letter to Pershing outlining what Patton thought were his singular attributes qualifying him for command of tanks. In the letter, Patton highlighted his cavalry experience because he thought working with light tanks would be "analogous to the duty performed by cavalry in normal wars." He explained that his previous experience as a machine-gun troop commander would prove valuable because it provided him with a working knowledge of the machine gun's mechanism and with skill in tactically employing the weapon, two skills he thought would be needed by a tank officer because "accurate fire is very necessary to good use of tanks."

Patton also pointed out his mechanical ability and French language capability, noting that "I speak and read French better than 95% of American officers . . . [and] I have also been to school in France and have always gotten on well with frenchmen." Finally, he stressed his aggressive spirit, adding that he thought he was "the only American who has ever made an attack in a motor vehicle."[21]

The motor vehicle "attack" to which Patton referred occurred on 14 May 1916, while he was serving as Pershing's aide de camp during the Punitive Expedition in Mexico. Patton and a party of fourteen men—including ten infantrymen, two civilian scouts, and two civilian chauffeurs—were sent to buy corn from nearby ranches for the American camp at Lake Itascate. After securing corn from a rancher at Coyote, Patton noticed about fifty men in neighboring Rubio. One of the civilian guides, E.L. Holmdahl, who had served under Villa, recognized several of the men as Villistas. Patton knew that one of Villa's chief lieutenants, Capt. Julio Cardenas, had family living at two nearby ranches and decided to search for him there.

Patton's plan of attack was simple: He would speed up, in the lead automobile, as he came within sight of the ranch, race past the house, and halt a short distance north of it. The other two cars were to halt just south of the house and dismount three men each to cover the southern end of the ranch. Six of the remaining men from these two cars were instructed to cover the road and the northern, western, and southern sides of the ranch house. The other two were to join Patton and his party on the east side of the house and assist in the search.

The plan was executed flawlessly. As the cars roared into position, three armed men fled from the house on horseback. Patton, the first into position, held his fire (Pershing had issued orders that Americans not shoot at Mexicans until certain they were hostile) and ordered the fleeing men to halt. They continued forward until they saw Patton's remaining men coming from the south, then shot at Patton from about twenty yards' distance before turning back. Patton returned fire, hitting Cardenas in the right arm and wounding his horse. The ensuing firefight lasted about fifteen minutes, and all three of the bandits were killed. Patton's force suffered no casualties.[22]

Maj. Robert Bacon, the headquarters commandant, endorsed Patton's letter and forwarded it to the AEF commander with the observation that he thought Patton was "unusually well equipped & fitted in every way for the command."[23]

Throughout the remainder of October 1917, however, Patton's ambitious nature caused him to waver in his resolve to join the Tank Service. The fact that the tanks were in an unsettled state caused him to seek command of an infantry battalion instead.[24]

In early November Patton changed his mind again and agreed to become the head of the AEF's Light Tank School, soon to be activated at Langres. He made this decision after Col. Paul B. Malone, director of the AEF schools, informed him that Colonel Eltinge had recommended him for the post, and after Major Bacon and Col. Martin C. Shallenberger, one of Pershing's aides, both advised him to throw in his lot with tanks. The reason was simple: An assignment with the fledgling Tank Service "aperes the way to high command if I make a go of it."[25]

Patton became the first American soldier officially assigned to duty with tanks when, on 10 November, AEF headquarters issued orders directing him to report to the commandant of Army schools at Langres for the purpose of establishing a tank school for the First Army.[26] First Lt. Elgin Braine, a Reserve artillery officer assigned to Battery D, 6th Field Artillery in the 1st Infantry Division, was ordered to report to Patton and serve as his assistant.[27] Braine, a highly trained technician, possibly a mechanical engineer, was well versed in the operation of internal combustion engines, blueprints, and other facets of industrial engineering. These skills made him an invaluable asset to the Tank Corps when a production program was later devised.

Whatever their individual strengths, Patton and Braine knew little or nothing about tanks, so, on 19 November, the two officers were ordered to report for two weeks of instruction at the French Light Tank Training Center at Chamlieu near Paris to prepare them for their duties.[28]

During the first week at Chamlieu, Patton had time to become thoroughly acquainted with the Renault tank. He drove the vehicle, noting its ease of handling and surprising comfort in contrast to the heavier British tanks. Although noisy, it could move at the pace of a running man, had a remarkably short turning radius, bucked and reared like a horse, and could easily bulldoze small trees. All of this greatly pleased the cavalryman Patton. The vehicle's only major drawback was poor visibility. When "buttoned up" (driving with all the hatches closed), the driver had only three small slits through which to observe the terrain in front of him. The gunner's visibility from the turret was little better.[29]

In addition to driving tanks, Patton fired their guns, observed a maneuver, worked on tactical problems, toured the repair shops and tank parking area, and spent long hours discussing how best to employ tanks in combat.

Braine joined Patton at Chamlieu on Tuesday, 27 November, for the second week of training. Patton thought this training was even more interesting than that conducted during the first week. Besides watching more maneuver training, Patton observed proficiency testing of drivers and gunners; translated lesson plans; saw tanks chained together in pairs for crossing wide, deep trenches; and was allowed to drive a Renault up and down steep banks. He also met several times with Gen. Jean E. Estienne, commander of the French tank forces, to discuss tank matters.[30]

On 20 November, while Patton was at Chamlieu, the British launched a major offensive at Cambrai. At 6 A.M., Maj. Gen. Hugh Elles kicked off the attack with a force of 350 heavy tanks. The British took the Germans completely by surprise, while a prearranged barrage fired by some 1,000 British guns added further to the Germans' confusion. The results of the attack were stunning: In just twelve hours the British advanced 10,000 yards from a base 13,000 yards wide, shattering two German divisions to their front and capturing 4,000 men and more than 100 guns. The British 3d and 4th Corps, which conducted the attack, suffered just 4,000 casualties. Compare this with the third Battle of Ypres, which required three months and appalling casualties to effect a similar advance.[31]

The Royal Tank Corps's performance at Cambrai helped to silence many of the tanks' critics, whose numbers had been growing, and vindicated the theories of Col. John F. C. Fuller, the Corps's operations officer.

Patton left Chamlieu on 1 December bound for Paris and the Renault Tank Works at Billancourt. While en route, he stopped at Albert with Col. Frank Parker to meet with Fuller. The trio discussed the attack on Cambrai, tank doctrine, and tactics.[32]

On 3 December, Patton and Braine toured the Renault factory. They were able to examine the light tank's design and construction and, during the course of their tour, recommended four minor improvements to the tank that the French later adopted. Their suggestions included a self-starter; a double-cased, felt-lined fuel tank that would prevent leakage if holed by enemy fire; an interchangeable mount so that each tank could carry either a 37mm gun or a machine gun; and a bulkhead between the crew compartment and the engine to protect the crew from fire.[33]

Patton was impressed during his tour of the Renault facility with the "great difficulty" the French tankers had in getting the manufacturer to cooperate. He perceived that American builders might be equally recalcitrant and, in his subsequent report on light tanks, included a veiled warning indirectly calling for officers charged with tank procurement to take a hard line when dealing with their civilian counterparts.[34]

Following their two-week orientation with the French, Patton and

Braine returned to GHQ at Chaumont and reported their findings orally to Colonel Eltinge, who was still temporarily in charge of the tank project for the AEF. They then set to work drafting a detailed report. In a letter to his wife on 5 December, Patton observed that his report was important "as no one knows any thing about the subject except me. I am certainly in on the ground floor. If they [the tanks] are a success I may have the chance I have always been looking for."[35]

Patton's double-spaced fifty-eight-page report was submitted on 12 December 1917. Later, when organizing his files, he penciled on the envelope containing the paper: "Original Tank Report. The Basis of the U.S. Tank Corps. Very Important. GSP."[36] Indeed it was. It served as the foundation for subsequent tank developments in the AEF, and at least one of his recommendations (a proposal that tanks be organized in platoons of five tanks, with three platoons to a company and three tank companies to a battalion) survived as part of American tank organization until the early 1980s.

The report is divided into four sections, including a detailed mechanical description of the Renault light tank, recommendations for the organization of tank units, a discussion of tank tactics and doctrinal theory, and proposed methods for the conduct of drill and instruction.

Patton described the light tank as a self-propelled armored vehicle capable of delivering predetermined firepower on the battlefield whenever needed. It had to be able to overcome all terrain obstacles in its path, provide maximum protection to both crew and engine, and be armed in order to accomplish this mission. He further specified that the vehicle must be easily manufactured in large numbers, have a favorable power-to-weight ratio, and be transportable to training or battle areas by either rail or truck.[37]

The Renault tank met all of these needs. It protected the crew from rifle and machine-gun fire as well as from the shrapnel produced by near misses from artillery fire. To protect a light tank from direct artillery hits would have required armor plating so thick as to make the tank's weight prohibitive, Patton said. The Renault's ability to maneuver on the battlefield with great agility combined with its small size (sixteen feet, five inches in length; five feet, eight inches in width; and seven feet, six and one-half inches in height) to make it difficult to spot, and thus reduced its vulnerability to large-caliber enemy guns.

The Renault's high degree of mobility was provided by its two caterpillar-type tracks mounted on the sides of the tank on frames called longerons. Each track ran endlessly around two large wheels, one of which propelled the track with power supplied by the engine while the other maintained proper tension and assisted the track's return. The vehicle's five-ton weight drove the lower portions of the tracks into the ground, thus providing traction. This weight was supported by a system of small rollers on the back of the lower parts of the tracks. The axles of these rollers were attached to a pair of rockers, called front and rear chariots, which were mounted inside the lower frames of the longerons. Power was supplied by an eighteen-horsepower, four-cylinder, monobloc, L-head motor that allowed the tank to attain a maximum speed of about seven miles per hour.

Armament consisted of either an 8mm machine gun or a 37mm gun mounted in a manually operated turret that permitted the gunner to engage targets through the full 360 degrees of the compass.

The remainder of this portion of Patton's report becomes highly technical and includes detailed specifications and mathematical formulas. There is little doubt that Lieutenant Braine contributed significantly to this section. Patton also described a number of proposed changes to improve the tank's performance in addition to the four mentioned earlier. Most of these changes were aimed at making the tank more comfortable for the larger American crewmen, providing for easier operation and maintenance, and facilitating recovery operations.

In Attachment B, Patton addressed the subject of light tank battalion organization. He proposed that platoons consist of an officer and fifteen enlisted men and five tanks — one equipped with a three-inch gun, two with six-pounder guns, and two with machine guns.

According to Patton, companies should have three platoons and a company headquarters element. The headquarters would have two officers and fifty-one enlisted men, including a first sergeant, supply sergeant, mess sergeant, signal (communications) sergeant, clerk, mechanic, ten drivers, twenty privates, two motorcyclists, ten chauffeurs, and three cooks. In addition to the fifteen tanks in the platoons, he suggested that the company have a tank equipped for signals, a tank for the company commander, eight supply tanks (which could also be employed for training or in reserve), five ammunition trucks, two trucks to carry petroleum and lubricants, a truck and trailer for baggage, a kitchen truck with trailer, an automobile, and two motorcycles. This plan would give each tank company five officers, ninety-six enlisted men, twenty-five tanks, and twelve wheeled vehicles.[38]

Battalions should consist of three tank companies, a headquarters section, and a repair unit. The headquarters and repair elements would have a combined total of three officers and forty-three enlisted men, including a major to command the battalion and two lieutenants to serve as the adjutant and the quartermaster. Two additional tanks would be provided for the battalion headquarters (one for the battalion commander and one for signals), plus three trucks, an automobile, and two motorcycles. The aggregate for the battalion would be eighteen officers and 331 enlisted men, seventy-seven tanks, and forty-two wheeled vehicles.

Patton further proposed that all of the supply trucks operate under battalion control to provide increased flexibility and that the company mechanics assemble periodically at battalion level to assist with major repair jobs. He also thought that the French repair and salvage system, which incorporated repair units at company and battalion levels and in the rear echelons, could be improved upon by establishing a large repair shop at some permanent or semipermanent location. Badly damaged tanks or wheeled vehicles or engines in need of complete overhaul could be transported to this center for repair.

Finally, Patton proposed that each battalion have an attached carrier company equipped with either seventy-seven trucks, each capable of carrying a tank in its bed or pulling one by trailer, or twenty-seven specially designed heavy trucks, each capable of carrying one tank and pulling another by trailer. He noted that the heavier vehicles would have to be limited to travel on only the best roads and concluded that the best solution would be a mix of both vehicle types. The basic principle in his mind was that adequate mobility would have to be balanced against cost and tonnage factors.

The remainder of this section of his report is devoted to a carefully thought out, detailed listing of the tools, spare parts, and special equipment that would have to be carried on each vehicle.

Patton opened his discussion on tactical matters with a brief history of tanks and lessons learned from their battlefield employment by the British and French. The critical part of this section is, however, his own recommendation for the employment of light tanks. He argued that the light tank should be seen as a "heavily armored infantry soldier" with "greater destructive and resistant powers," and not as artillery or an independent combat arm. Patton said the main purpose of light tanks was to assist the infantry in breaking through the enemy's forward positions. Once this task was accomplished, the tanks could then "assume the role of pursuit cavalry and 'ride the enemy to death.'" This latter possibility led him to recommend that the light tanks be deployed in depth for shock effect, and that a substantial reserve force be maintained in order to exploit a breakthrough.[39]

Patton envisioned several missions for the tanks in their infantry support role. First, they must clear wire obstacles for the advancing doughboys. Second, they would have to employ their weapons in a manner that would prevent the enemy's infantry from manning the trench parapets after the preparatory artillery barrage lifted. Third, they would be required to suppress enemy machine guns and trench cannon. Fourth, the tanks would be required to assist the infantry in mopping up the objective by masking strong points and blockhouses with fire and smoke bombs, remaining on the objective until the infantry had consolidated their gains. Fifth, tanks would guard against enemy counterattack by patrolling throughout the sector between the most advanced infantry positions and the rear of friendly artillery fires. Finally, acting on their own initiative, the assault tanks (joined by their supporting elements and possibly the reserve) would press on to exploit the attack, seeking "every opportunity to become pursuit cavalry."[40]

To accomplish these tasks, Patton contended that the tanks must be deployed in depth, with their sectors of attack corresponding with those of the infantry. He thought that his proposed organization would ideally suit this plan, allowing one tank platoon to accompany each attacking infantry company and a tank company to support each infantry battalion.

To prevent the misuse of tanks when both light and heavy tanks were employed in the same sector, Patton suggested that the division commander exercise tactical control "because of the danger that the commander of whichever group was the senior might use the tanks of the other variety to the advancement of his own tanks and the disadvantage of the other."[41]

Patton recognized the importance of reconnaissance and recommended that company commanders and platoon leaders be given the opportunity to view from the air the terrain over which they would be required to maneuver. He also proposed that leaders be required to physically walk the routes from their assembly areas to the line of departure by day and night in order to ensure their timely arrival at zero hour. This, he said, would prevent losses to enemy artillery fire should the tanks arrive too early and be forced to remain stationary.

Patton concluded with the observation that heavy tanks were more independent and should thus precede light tanks in the attack—especially when no artillery preparation was employed—capitalizing on the heavy tanks' superior ability to cut wire. Nevertheless, he thought light tanks held an advantage in mobility because they could be easily transported by truck or trailer, while the heavy tanks could only be moved by rail.[42]

Subsequent operations proved Patton's theories to be at least partially correct. Employed organizationally and tactically almost exactly as he had proposed, the light tanks had a difficult time keeping pace with the infantry during the St. Mihiel offensive because of poor ground conditions and the rapidity of the German retreat. They were, however, a valuable asset in support of the Meuse-Argonne offensive. His ideas on battlefield mobility were never tested because the AEF lacked sufficient trucks to transport the light tank force. Instead, the tanks had to be moved by rail to both the St. Mihiel and the Meuse-Argonne sectors, then under their own power during operations. This meant conducting long road marches to link up with units at the front, contributing to a high rate of mechanical failure as the tanks were forced to conduct extended operations without the benefit of overhaul.

In the final section of his report, Patton formulated a plan "recommended as best suited for the quick production" of sufficient numbers of tank personnel to man the proposed force. To carry out his plan, he advocated the establishment of separate tank centers and schools for the light and heavy tanks, with the schools' commandants also acting as the center commanders.[43]

Working on the assumption that tanks would be made quickly available in sufficient numbers, Patton proposed that enough officers and men be recruited and shipped to each tank center to fill a company plus provide for one or two additional instructors. He thought these men should have either experience driving automobiles or motorcycles or mechanical skills derived from work as blacksmiths, foundry hands, plumbers, or gas fitters.

Upon arrival at the center, the men would undergo a four-week training program. They, in turn, would act as instructors for a second company arriving at the beginning of the fifth week. Within two months these two companies could then train an additional two to four more, "giving a total of six companies or two battalions instructed at the end of three months."[44]

Patton claimed that this system, in addition to turning out fully trained units, would give the fledgling tankers the opportunity to practice the principles just learned. It would also enhance unit cohesiveness by allowing the officers and men to become accustomed "to each other and to each other's methods."[45]

His second proposal was more realistic, assuming as it did that men and not machines would be available in quantity for training. For such a program he recommended that as many officers and noncommissioned officers as possible be recruited and sent to the centers. The men would be trained under the same system but with six men assigned to each vehicle. Patton proposed that graduates be sent to other schools — such as those for machine guns or small cannon — if sufficient tanks were not yet available to create companies at the end of the four-week training cycle, holding over only the best qualified graduates to serve as assistant instructors for the next batch of students arriving at the tank center. While less efficient than the unit-training proposal, this plan was eventually employed because of the difficulty the Tank Corps experienced obtaining vehicles.

Patton had been less than thrilled by the performance of Reserve officers called up to help fill the ranks of the rapidly expanding Army. Because of this, he recommended that "at least one regular officer per company [be detailed] to enforce discipline, as it is a well-known fact that working with machines has a very disastrous effect upon discipline. It seems to run out of men as the oil soaks into them."[46] Patton also proposed that soldiers who showed marked mechanical ability be given additional training in repair and salvage so they might later be employed as mechanics in battalion repair shops.

The program of instruction Patton described stressed driving skills and marksmanship training, beginning with the basics and making the training progressively more difficult. Routine crew maintenance procedures would also be emphasized. Lectures would be devoted to the theory of gasoline engines and other vehicle systems; the tactics, history, and employment of the tank in combat; drill; and signaling techniques. Students would be required to demonstrate their ability to maneuver the tank, fire its weapons, and respond to visual signals during a comprehensive examination administered at the end of the course.

Patton noted that the proposed course was far from all-inclusive. Subjects such as map reading, intelligence gathering, and camouflage were not included in his proposal because he thought these skills could best be taught after the men joined their units.

Finally, to save time, Patton suggested that the course could be shortened to three weeks for selected officers and men, who could then receive training in the omitted subjects during the fourth week or in the evenings.

When Patton finished, he was pleased with his efforts. He wrote to his wife two days after submitting the report, noting that he had been required to account for "all conceivable conditions" and to plan to the minutest details, providing "a list of tools and spare parts down to and including extra wire and string" and the numbers and ranks of soldiers in units. He had, he wrote, accomplished all that with "nothing to base it on but a general knowledge of soldiering." In a self-congratulatory mood, he added that he did not "think many men could have combined the exact mechanical knowledge with the general Tactical and organizational knowledge to do it. But I think I did a good job. Infact I surprised my self and hope others will think as well of it as I do."[47]

Others did. Patton's work was so thorough, his proposals so well reasoned, that the majority of them were enacted. He had ample opportunity to put them to the test as commander of the first tank center to be established – the light tank center and school at Langres. The relative ease with which he was able to recruit and train a full brigade of tank troops for combat in less than five months bears witness to the soundness of his ideas.

Rumors began circulating in early December 1917 that Col. Samuel D. Rockenbach, a cavalry officer detailed to the Quartermaster Corps, would soon be appointed to command both the light and heavy tanks in the AEF. Patton, who had grown somewhat discouraged in the face of the awesome task awaiting him, told his wife, "I hope I can make a success of this business but starting with nothing is hard. After we get a little nucleus it will go easier. Now I feel helpless and almost beaten but I will make a go of it or bust[,] Rockenbach or no."[48]

Similar thoughts must have occurred to Colonel Rockenbach the morning of 22 December 1917, when he reported to Colonel Eltinge at GHQ and announced that he had received orders appointing him chief of the Tank Corps. Eltinge, who had been supervising tank matters for Pershing's staff to that point, "pulled open the lowest drawer of his desk, took out a bundle of papers, [handed them to Rockenbach] and said, 'Here's all we know about Tanks, go after them.'"[49]

Notes

1. Blumenson, *Patton Papers,* 437; Timothy K. Nenninger, "The World War I Experience," *Armor* 78, no. 1 (Jan.–Feb. 1969): 47.
2. Rockenbach, "Tank Corps Operations," 1.
3. Brig. Gen. Samuel D. Rockenbach, "The Tank Corps: A Talk to the General Staff," 24 Sept. 1919, USAMHI, 1; Nenninger, "The WWI Experience," 47.
4. Rockenbach, "Tank Corps Operations," 2.
5. Ibid., 3.
6. Ibid.
7. Cable No. 159-S, para. 15, 14 Sept. 1917, included as an enclosure to Lt. Col. Robert L. Collins, "Report on the Development of the Tank Corps," undated, Modern Military Records Division, Record Group 120, Entry 22, Folder 387, National Archives, Washington, D.C. (hereafter NA).
8. Ibid.
9. Ibid.
10. Rockenbach, "The Tank Corps," 4.
11. Rockenbach, "Tank Corps Operations," 2.
12. Majs. James A. Drain and Herbert W. Alden, "Report of Investigations by Majors Drain and Alden," 10 Nov. 1917, included as an enclosure to Maj. R. E. Carlson, "Memorandum on the Development of Tanks," 16 Mar. 1921, USAMHI, 10.
13. Ibid., 11.
14. Ibid., 7–8.
15. Ibid., 2.
16. Ibid., 8.
17. Ibid., 9.
18. Rockenbach, "Tank Corps Operations," 2.
19. George S. Patton, Jr., to Beatrice Ayer Patton, 19 Sept. 1917, as quoted in Blumenson, *Patton Papers,* 423. Following Blumenson's example, I have elected not to use [*sic*] to identify Patton's misspellings or punctuation and grammatical errors. In some instances I have altered punctuation or grammar to clarify a passage for the reader. Patton was a notoriously poor speller, and his writings include numerous grammatical errors.
20. Undated Patton diary entry as quoted ibid., 426.
21. Capt. George S. Patton, Jr., to commander in chief, AEF, through headquarters commandant, AEF, 3 Oct. 1917, subject: Command in the Tank Service, as quoted ibid., 427.
22. Blumenson, *Patton Papers,* 359–66.
23. Ibid., 427.

24. George S. Patton, Jr., to Beatrice Ayer Patton, 9 Oct. 1917, as quoted ibid., 428–29; Patton diary entry, 3 Nov. 1917, as quoted ibid., 432.
25. Patton diary entry, 4 Nov. 1917, as quoted ibid., 432.
26. Capt. George S. Patton, Jr., report, "Subject: Light Tanks," 12 Dec. 1917, George S. Patton, Jr., Collection, Patton Chronological File, 11–12 Dec. 1917, Library of Congress, Washington, D.C. (hereafter Patton Collection); Col. George S. Patton, Jr., "History of the 304th (1st) Brigade Tank Corps," undated, Patton Writings, Box 49, Military Writings 1918–20 File, Patton Collection, 1.
27. Capt. Elgin Braine, "Personal Experience Report," 22 Dec. 1918, Patton Military Papers, Box 47, Personal Experience Reports of Tank Operations–1918 File, Patton Collection.
28. Patton, "304th Brigade History," 1; Patton, "Light Tanks," 1.
29. Blumenson, *Patton Papers*, 444.
30. Ibid., 445.
31. Jones et al., *The Fighting Tanks*, 21, 25.
32. Blumenson, *Patton Papers*, 446.
33. Patton, "Light Tanks," 1; Rockenbach, "Tank Corps Operations," 4.
34. Patton, "Light Tanks," 1.
35. Blumenson, *Patton Papers*, 447.
36. Patton, "Light Tanks," 1.
37. Ibid., "Attachment A, Mechanical."
38. Ibid., "Attachment B, Organization."
39. Ibid., "Attachment C, Tactical."
40. Ibid.
41. Ibid.
42. Ibid.
43. Ibid., "Attachment D, Instruction."
44. Ibid.
45. Ibid.
46. Ibid.
47. George S. Patton, Jr., to Beatrice Ayer Patton, 14 Dec. 1917, as quoted in Blumenson, *Patton Papers*, 457.
48. Ibid., 457–58.
49. Rockenbach, "The Tank Corps," 3.

CHAPTER II

The Light Tanks:
From Langres to St. Mihiel

On Christmas Eve 1917 the AEF Tank Corps was a woeful force indeed, consisting as it did of just three officers: Colonel Rockenbach, Captain Patton, and First Lieutenant Braine. The task facing them was monumental. They would be required to adapt a new piece of machinery for use on the battlefield, work out designs and arrange for production, establish tables of organization and equipment for tank units that would be suited to the existing organization of the United States Army, work out tactics suitable for the employment of tanks, establish needed coordination measures to ensure cooperation among the various combat arms, recruit sufficient officers and men to man the force, and devise a detailed training program designed to turn out tank crews and units ready for combat.

Patton had high hopes that Rockenbach would be able to chart a straight course for the Tank Corps, although he had misgivings as to how their personal relationship would develop. The day he reported to Rockenbach for duty, Patton wrote his wife:

> The whole Tank program is in a mess now as about three departments are trying to run it but Col Rockenbach will I hope straiten that out if he does nothing else. I guess he does not care a whole lot for me but my theory that if you do your best no one can hurt you will be put to the proof.[1]

Patton need not have feared about the relationship. His chief biographer, Martin Blumenson, observed that although neither officer developed an initial liking for the other, the older Rockenbach brought a wealth of experience and maturity to his post as commander of the tank force and principal tank adviser to General Pershing. Hardly an original thinker, even-tempered, lacking a sense of humor, and displaying a tendency to fixed, narrow opinions, Rockenbach was able to balance Patton's headstrong enthusiasm and channel his creativity.

According to Blumenson, Patton and Rockenbach, who recognized that their future success depended on their ability to work together, were able to cooperate in the interests of the war effort and their own personal ambitions. The pair shared family roots in Virginia, and both had attended the Virginia Military Institute (Patton in 1903 and 1904 before going on to West Point; Rockenbach graduating in 1889). They soon discovered that their personalities and character strengths were complementary. While never becoming a close friend, Rockenbach eventually became one of Patton's admirers and strongest supporters – though Patton always remained uncomfortable with his superior.[2]

Patton's first task as director of the Light Tank School at Langres was to find a suitable location in the area at which to establish the school. Patton found what he thought was the perfect place while scouting the Langres vicinity with Lieutenant Braine in late December. The site was at Bourg, some five miles south of Langres on the Dijon road. It was a rising piece of ground in the Bois d'Amour and was flanked by two good roads and a railroad. Billeting could be conveniently obtained in Bourg and the neighboring villages of St. Geosmes and Brennes and, if necessary, in Langres.

Rockenbach approved Patton's proposal to obtain the land on 29 December and returned to GHQ in Chaumont to requisition it. Braine accompanied him, with the mission of securing buildings for classrooms, offices, and billets.[3]

In early January 1918 Patton escorted Rockenbach to the French tank training center at Chamlieu and the British Tank Corps headquarters at Bernecourt. During the trip, designed to acquaint Rockenbach with British and French tank tactics, equipment, and training, the pair met with Gen. J. E. Estienne of the French tank service, Maj. Gen. Hugh Elles of the British Royal Tank Corps, and Col. John F. C. Fuller, Elles's chief of staff.[4]

The first increment of officers to be assigned to the Tank Corps arrived on 8 January. These ten men had all been commissioned from the enlisted ranks and were made available by the Coast Artillery Corps. Because the tank school site still had not been approved, Braine escorted the officers to Langres, where half were sent to the 37mm gun school and half to the machine-gun course. Two days later, two of the officers, 2d Lts. Edward A. Redmon and Marion A. Friend, were ordered to be discharged for medical reasons. They left on 15 January.[5]

Word was received during this period that the French did not want to give up the ground Patton sought for his tank school. They proposed the selection of an alternate site. This infuriated Patton, who, in a fit of pique, called them "the d—fools."

Meeting to discuss the issue with a French colonel, Patton told him "that if he would do a little more to help" they might be able to get along better. He also suggested to the officer that "the reason the war was lasting so long was that [the French] were too afraid of civilians." The colonel responded to this tirade with the observation that the ground would be of little value to the Americans because they would be unable to conduct live-fire gunnery training. Patton told the colonel just to get him the ground and he "would attend to that, for I fully intend to shoot[,] French or no French."[6]

After submitting a renewed request for the ground at Bourg, Patton spent two days working with Braine on spare-parts requirements for the tanks, observing that "it takes a lot of stuff to kill a German."[7] Patton waited until the spare-parts lists were typed, then delivered them personally to Rockenbach in Chaumont on 15 January. Patton's report consisted of twenty-four pages of detailed tables listing the material needs for a light tank battalion and its supporting repair unit—right down to the number of screws and bolts that should be kept on hand. Patton and Braine having done the stubby pencil work, it would be up to Rockenbach to use his seasoned staff skills to get what was needed.[8]

In addition to helping Patton with the spare-parts requirements, Braine had acquired office space, coordinated with other AEF schools to train newly arrived tankers, organized a program of instruction for a mechanical course, and obtained a broken-down Atlas truck and surplus tools from the Quartermaster Corps for training purposes.[9]

On 18 January a second contingent of eleven Coast Artillery officers arrived in Langres. Like their predecessors, these officers had also been commissioned from the enlisted ranks. Braine sent all but 2d Lt. Thomas C. Brown off to the machine-gun and small-cannon courses. Patton had selected Brown to serve as adjutant of the tank school and busied him with the task of organizing a headquarters.[10]

Patton's frustrations over the land acquisition issue were finally relieved on 17 January, when, after consultation with officers at the French Mission, he was able to secure an agreement giving him the ground he wanted at Bourg and permission to establish firing ranges.[11]

Two infantry officers, 1st Lt. Loyall F. Sewall and 2d Lt. Horace C. Nelms, reported for duty on 24 January, after having served at the front in the Toul sector with the 1st Infantry Division's 18th Infantry Regiment. Sewall and Nelms were given duties to perform in Langres until the twenty-ninth, when the other officers completed their gunnery training. Sewall and Nelms then embarked together on a course of instruction that included a special ten-day mapmaking and -reading class taught at the Army School of the Line; lectures on the use of camouflage, gas defense, and aerial photoreconnaissance; and mechanical training on the engine and transmission of the Atlas truck conducted by Lieutenants Braine, Ellis Baldwin, and Fred C. Winters.[12]

While Patton's officers were receiving this training, he worked on a lecture he delivered to the General Staff at Chaumont on 23 January, smoothed over a bureaucratic blunder that almost resulted in the Americans' refusing to accept the land he had worked so hard to get from the French, and gave tours of the new tank school grounds to senior American officers and a pair of visiting French officers from the tank center at Martigny-les-Bains.[13] He also busied himself with paperwork, disciplinary matters, and fretting over another problem—the lack of tanks:

> Unless I get some Tanks soon I will go crazy for I have done nothing of any use since november and it is getting on my nerves.
> I cussed a reserve officer for saluting me with his hands in his pockets to day and he said that he demanded to be treated like an officer. I almost hit him but compromised by taking him to the General [probably Maj. Gen. James W. McAndrew, commandant of the AEF schools] who cussed him good. Some of these new officers are the end of the limit. I bet the Tank Corps will have discipline if nothing else.[14]

Patton's emphasis on discipline was far from popular with his officers, and he told his wife that "I am getting a hell of a reputation for a skunk. . . . I expect some of them [Reserve officers] would like to poison me. I will have to eat [only] eggs like Louis XI."[15]

With the acquisition of the land near Bourg approved, Patton knew it would be a matter of weeks before enlisted men began arriving to fill the ranks of the training companies. With this in mind, he published his first directive for school personnel. Dated 27 January, the document specifies rules for appearance and deportment, with particular attention being devoted to the proper rendering of the hand salute. In the directive Patton instructed officers to obtain leather and brass polish and use it to keep shoes, belts, leggings, boots, belt buckles, and other metal accouterments "brightly shined." He also ordered the men to use gasoline or some other cleaning material to keep their uniforms free of spots and stains. Hair was to be kept short enough so that the officers and men of the tank school would "look like soldiers and not like poets." Finally, Patton told the officers that he would hold them personally responsible for the appearance and actions of their men.[16]

Much of Patton's fixation on matters of discipline, appearance, and deportment may have stemmed from the fact that he had little else to occupy him during this period. He noted in a letter to his wife that it was difficult for him to keep everyone busy while they awaited the arrival of their first tanks. Although pleased by a compliment paid him by Major General McAndrew, who told Col. Hugh A. Drum that Patton "was taking hold better than any man he had," Patton still was not personally satisfied:

> If I ever do feel that I am earning my pay I may realy begin to get some where. I sort of wish I had gotten a battalion of infantry then I could have seen some results where as it is all so far in the future that one can't see it but in the minds eye.[17]

At the end of January Patton secured permission from Rockenbach to detach Lieutenant Braine and send him to the United States to act as the liaison between the Tank Corps in France and manufacturers tasked with building light tanks in America. Patton considered this a major accomplishment, as he still found his working relationship with Rockenbach less than congenial. Patton noted in his diary that getting Rockenbach to agree to the transfer "was hard but finally he took the hook like a fish."[18] He also told his wife that he thought the Tank Corps's chief was "the most contrary old cuss I ever worked with. As soon as you suggest any thing he opposes but after about an hours argument comes round and proposes the same thing him self. So in the long run I get my way, but at a great waste of breath."[19]

On 1 February, the day Braine departed for Paris, Patton unveiled an innovation he had worked out to train his crews for combat despite the lack of vehicles—"machine foot drills." These drills were designed to conform to the Infantry Drill Regulations yet accustom the crews to the various signaling methods and battle drills they would employ while maneuvering their tanks in combat formations.[20]

The drills Patton described were to be initiated upon commands transmitted visually (using flags), by touch (tank commander to driver), orally, with a Klaxon (horn), or by the example of leaders. For example, to bring his platoon into a line formation, a platoon leader would move a red flag several times from right to left and left to right. To halt, he would first press the top of his driver's head, indicating that the driver should stop. The other vehicle commanders, seeing the platoon leader and his driver halt, would take that as their cue to do likewise. Numerous other drills were practiced in similar fashion, allowing the crews to become familiar with vehicle intervals, commands, and platoon and company combat formations. Although the specific drills and signals have changed with the years, some tank units in the United States Army today still employ the technique to practice platoon and company battle drills when funding constraints restrict the amount of time they can spend maneuvering their tanks.

In addition to training, Patton put his officers to work preparing barracks and workshops and installing water lines in anticipation of the arrival of the first draft of enlisted volunteers from other AEF units. While much of this work was accomplished with materials obtained through proper channels, some "midnight requisitions" were needed. Patton recalled that six buildings mysteriously "sprang up in the night like mushrooms." They looked suspiciously like French barracks but had been "freely daubed with serial numbers . . . backed with [a] large obtrusive U.S. A still closer examination would have revealed places where other numbers had been removed with a plane. But why be inquisitive."[21]

Patton, ever rank-conscious, learned from a visiting West Point classmate that he had been promoted to major on 15 December. This did not constitute official notice, but after considerable internal debate, Patton decided to pin on the insignia of the higher grade[22] and instructed his adjutant to prepare correspondence with a signature block reading "G.S. Patton, Jr.; Maj. of Cav.; Commanding."[23] In reality, however, his date of rank as major was 26 January, and orders would not arrive in France for another month and a half.

On Sunday, 17 February 1918, Lt. Col. F. B. Hennessey and 2d Lt. Will

G. Robinson arrived with a replacement draft of 200 men from artillery units in the 42d Infantry Division.[24] These men would become the nucleus of Patton's light tank brigade. Included in their ranks was Sgt. Maj. George B. Heilner, who recalled that they were initially billeted in St. Geosmes, then moved to Bourg on the twenty-second to begin training. Hennessey, because of his seniority, was returned to his unit.[25]

With the arrival of the men, Patton was able to organize two companies under the command of Lieutenants Williams and English. Platoon leaders were also assigned and officers appointed to teach classes in mechanics, driving, weapons, and athletics.[26]

The training day during this initial phase of unit training began at 8:20 A.M. with thirty minutes of close-order drill. This was followed by a half hour of athletics; a half hour of platoon signal drill; a thirty-minute lecture on subjects such as guard duty, military courtesy, posting sentinels, and challenging; and a half hour of platoon machine foot drill. In the afternoon all officers (except the duty officer) and ten noncommissioned officers (NCOs) from each company received forty-five minutes of pistol instruction, conducted forty minutes of company-level machine foot drills, and underwent mechanical training for an hour and a half. The enlisted men performed housekeeping and other chores during this period.[27]

Reflecting on this training regimen in a letter to his wife, Patton noted that he thought it would "keep them fairly on the go for a while till we get the tanks. It is absolutely necessary to do so as there is nothing else to amuse them." He abhorred the idea of laxness and added that he thought "our chief fault as an army has been that of taking things too easily. When the days get longer I can work them more which will be a help to them."[28]

The pressure to train tankers was mounting, and Patton rebelled at a proposal by Rockenbach calling for a total of forty trained light tank companies by 30 June. The schedule was predicated on the shipment of a total of 930 tanks during that period.

Patton outlined his concerns in a detailed memo to Rockenbach on 2 March. Chief among them was his expectation that the fledgling supply system would fail. "You know what happened in the case of our trucks," he wrote. "Bodies landed in one place. Chassis another and both were useless. No spare parts at all. Tanks would be an even worse case [because the quartermasters would not know what to do with them]."[29]

To alleviate this problem, Patton suggested paring the number of companies to be trained to twenty-eight and recommended that an officer from the Tank Corps "with rank and experience" be detailed to act as Rockenbach's representative in the United States to coordinate the shipment of men and tanks.

Rockenbach disapproved this plan, much to Patton's dismay. Someone in the War Department must have been thinking along the same lines as Patton, however, because on 5 March the secretary of war appointed Lt. Col. Ira C. Welborn, an infantry officer awarded the Medal of Honor during the Spanish-American War, to serve as director of the Tank Service in the United States. The AEF Tank Corps and the Tank Corps in the United States would remain two separate entities. Welborn's role would be largely administrative: supervising the recruiting and initial training of Tank Corps personnel, overseeing their preparation for overseas movement, and working with the Ordnance Department on tank production.[30]

With the arrival of the enlisted men, Patton began actively recruiting additional officers for the Tank Corps. Among those he selected were 1st Lts. Harry H. Semmes and Newell P. Weed, both of whom reported for duty in early March. Weed recalled that they were required to take a "test of suitability," which consisted of an examination on gas engines. Weed said they prepared for the test by reading a small book "issued by some oil company." Both passed, though neither knew "a magneto from a carburetor."[31]

The Light Tank School was able to obtain two Hotchkiss 8mm machine guns and two 37mm cannons from the Army schools at Langres, and Patton immediately introduced instruction on these weapons into the schedule after 1 March.

Patton also designed a course on "visual training." This incorporated instruction on both ground and air observation techniques, and he was able to supplement it with aerial photographs of the Bourg area obtained from the Signal Corps. Patton also instructed Sergeant Major Heilner to use the photos to update the maps of the area for the map-reading classes.[32]

While making his rounds of the other Army schools, Patton encountered Capt. Sereno E. Brett, an instructor in the Infantry Specialists School who was an expert with the 37mm cannon. Brett was anxious to join the tanks and impressed Patton. Unable to secure Brett's release from the school's commandant, Patton petitioned Rockenbach to request the services of this "invaluable" officer, as that was the only way the transfer would be approved.[33] Rockenbach concurred, and Brett, who was to play a major role in subsequent events, reported for duty on 28 March to serve as Patton's senior instructor.[34]

Personnel and training affairs all were in order, but "a Tank Corps without tanks is quite as exciting as a dance without girls."[35] The picture brightened considerably on 23 March, when ten Renaults, long ago promised by the French, arrived by train at Bourg. Patton arranged with the railroad to move the train opposite the Bois D'Amour, where he had previously ordered unloading platforms constructed. He assisted personally with the operation, leading nine of the most experienced drivers on a mile-long journey from the rail line to the tank park.[36]

Patton recalled later that

> the french were much impressed as they said it would take us 15 hours and we did it in just three.
>
> It was a beautiful moonlight night and ideal for the purpose. When the procession of ten started across the fields I was delighted as I have been living on hopes for the last four months.
>
> They certainly are saucy looking little fellows and very active. Just like insects from under a wooden log in the forest.
>
> No one but me had ever driven one so I had to back them all off the train but then I put some men on them and they went along all right. I took one through some heavey woods this morning and it just ate up the brush like nothing.
>
> Tomorrow they will all be oiled up and we will start active business making drivers on Tuesday. It will be fine haveing something to work with as up to now I have had one old truck [the Atlas] . . . and had to teach all about driving with it.[37]

Sufficient progress had been made with the truck, however, that a program of tank-driving instruction was initiated the day after the tanks arrived. Ten men who had shown "marked ability" were initially trained to serve as instructors, and then details were selected from each of the companies to undergo training.[38] Patton's goal was to qualify ninety-six men and three officers as both drivers and gunners within three weeks.[39]

Drivers found the learning experience to be a painful one. Tank commanders were required to transmit commands to their drivers by kicking them. This was the only means of internal communication, as the Renaults lacked a radio intercom system and were too noisy for voice commands to be heard. To get the driver to move forward, the commander kicked him in the back. Similarly, a kick to either shoulder signaled a turn in the direction of the shoulder kicked. The signal to stop was a kick to the driver's head, while repeated kicks to the head meant the driver should back up.[40]

By 26 March Patton was able to observe that training was going "full blast" and that he finally felt like he was accomplishing something. Although he had nowhere near the number of tanks needed (or promised to him), the ten he had were "much better than none." He was also impressed by the speed with which his soldiers were learning to drive the tanks and noted that he thought Americans could be taught the techniques "50% faster than the French."[41]

Driving the Renault was indeed remarkably easy. The driver's controls consisted of a clutch pedal on the left of the floor, an accelerator pedal in the center, and a parking brake pedal on the right. The engine was started by means of a hand crank located at the back of the gunner's compartment on the firewall separating the gunner from the engine compartment. The driver could control the vehicle's speed by either depressing the accelerator pedal or using a hand throttle control located on the right side of the driver's compartment. A spark control lever was also provided, allowing the driver to advance or retard the ignition spark, depending on the amount of strain on the engine. Two large steering levers, one on each side of the driver's seat, acted as the service brakes when pulled simultaneously. To steer to the right, the driver merely pulled back on the right lever, braking the track on that side of the tank. The left-side track would continue moving at normal speed, pivoting the vehicle to the right. A similar procedure was used to turn to the left.[42]

The most difficult task for drivers to master was negotiating short, extremely steep grades. The trick was to learn to slip the clutches in such a way as to allow the vehicle to return to the horizontal gently, without a crashing jolt, as it cleared the top of the obstacle. For this, experience proved to be the best teacher.[43]

To make mechanical training even more effective, Patton asked Rockenbach to provide the tank school with a number of visual aids, including a complete light tank engine cut to show the working internal parts such as valves, pistons, and the camshaft; a carburetor cut to show the working of the float, needle valve, and jets; a clutch cut to show the working of its internal parts; and other mechanical assemblies.[44]

By the end of March Patton had a sizable force at Bourg, and, with the prospect of even larger replacement drafts due in, he was forced to seek additional space for housing and training them. He found it in Brennes, a small neighboring town located within twenty minutes' marching time of a proposed railroad siding in the Bois St.-George-le-Gros.[45]

On 5 April ninety men arrived and were organized into Company C under the command of Second Lieutenant Sledge. Additional replacements began arriving at the rate of about ten per day as units responded to an AEF directive authorizing the transfer of "suitable" volunteers to the Tank Corps. Patton's task of going to units in search of qualified officers was made easier by an order for all officer applicants to report to Bourg for screening. "As only about one-third of those applying were accepted, a very good lot were obtained."[46]

Following the approval in late March of a revised table of organization for the AEF Tank Corps calling for five battalions of heavy tanks; twenty battalions of light tanks; and the necessary headquarters elements, repair and salvage companies, depot companies, training centers, and replacement companies, Rockenbach issued an order providing for numerical designations for the various units.[47] Light tank battalions were to be numbered 1 to 40 and heavy battalions 41 to 50 in order of organization. Companies would be designated A, B, and C. Tank Centers were to be numbered 1st, 2d, and so on as they were formed. Repair and salvage companies would be allocated to each center and would take the numerical designation of their respective centers, while training and replacement companies for light tanks would be numbered 1st, 2d, and so on, and for heavy tanks, 41st, 42d, and so on.[48]

Patton, who had only recently received written confirmation of his promotion to major, was notified on 1 April that he had been confirmed as a lieutenant colonel.

By mid-April the level of training of the officers and men at the 1st Tank Center had reached a point at which Patton thought he could begin conducting maneuvers. On 16 April he initiated the first of a series of simulated combat exercises, maneuvers, and practice movements for which he personally wrote and supervised the preparation of field orders, instructions, movement directives, and other details designed to encompass a wide variety of problems and combat situations.[49]

One of the highlights of this period was a tank demonstration performed for the officers of the General Staff College on Monday, 22 April.[50] The demonstration featured Patton's ten light tanks in support of a composite battalion of infantrymen from the Army schools at Langres. Two battalions of artillery also participated. Patton was especially pleased because the demonstration provided not only an excellent training vehicle for his own men but also a chance to impress the students and faculty at the staff college. He later wrote that he thought the maneuver played an important role "in convincing a great number of the staff officers who witnessed it, of the efficiency of the light tank."[51]

The 1st Light Tank Battalion was organized on 28 April with Patton in command. Company A was given to Capt. Joseph W. Viner, a cavalryman who had joined the Tank Corps the previous month after arriving in France following an abbreviated tour as a mathematics instructor at West Point. Brett took over Company B, and a third Regular Army officer, a Captain Herman, was given command of Company C.[52]

In addition to realistic, demanding training, Patton stressed strict discipline and the maintenance of a high level of morale. At one point Patton, inspired by soldiers of the 82d Infantry Division he had seen sporting shoulder patches, challenged the officers of the tank center to design a similar patch. "I want you officers to devote one evening to something constructive," 2d Lt. Will G. Robinson later quoted Patton. "I want a shoulder insignia. We claim to have the firepower of artillery, the mobility of cavalry, and the ability to hold ground of the infantry, so whatever you come up with it must have red, yellow, and blue in it."[53]

Robinson and his roommate spent the rest of the evening with crayons "I had liberated in front of the fire place" figuring out a design. They decided on a "pyramid of power, but had a hell of a time" dividing it into the three colors. They finally drew three lines terminating in the center. The next morning, Patton adopted their design; he gave Robinson a $100 bill and sent him into Langres to get as many patches as possible made by Retreat. Robinson bought felt in the three colors and went to a hat-and-cap shop to have the patches sewn. He returned to camp with about 300 patches and time to spare. "Patton was tickled about it. If there was anything he wanted, it was to make the Tank Corps tougher than the Marines and more spectacular than the Matterhorn. That triangle was the first step."[54] (The triangular patch of the AEF Tank Corps was later embellished with an endless track, a cannon, and a lightning bolt superimposed over the three colors, and it remains the shoulder patch used to denote armored divisions.) Patton sought to further differentiate his tankers from other soldiers by designing special collar insignia for the Tank Corps.[55] The collar insignia consisted of a silhouette of a British Mark IV heavy tank set within a wreath. This was embossed on a circular brass disk for enlisted personnel. The officers' collar insignia was designed without the brass backing. Buttons featuring the same design were worn on the uniform coat.

If other officers thought Patton was promoting the Tank Corps as an independent arm or sought to usurp infantry missions, he allayed their fears in a lecture to officers at the School of the Line by reminding them that "tanks in common with all other auxiliary arms are but a means of aiding infantry, on whom the fate of battle ever rests, to drive their bayonets into the bellies of the enemy."[56]

Patton met with each group of newly assigned officers and enlisted men and impressed on them the need for strict discipline. Discipline, as Patton described it, was "instant, cheerful and automatic obedience."[57] He likened the discipline required of soldiers to that exhibited by players on a football team. He compared officers to quarterbacks and enlisted men to the other players on the team, noting that their obedience must be instant and without thought because, while failure on the football field might result in the loss of a few yards, the "lack of discipline in war means death or defeat, which is worse than death. The prize for a game is nothing. The prize for this war is the greatest of all prizes—Freedom."[58]

After dismissing the enlisted men, Patton told the officers that everything they had just heard him tell the soldiers "applies in treble force to yourselves." He was especially critical of the democratic approach he had heard officers take when issuing orders. Continuing the football analogy, he told them:

> Do you ever hear the quarter-back of a football team say "Please give me your attention gentlemen, the signal which I am about to call is ——,["] You never did. There is nothing harsh in the brief words of command, no more than there is impoliteness in the brief wording of a telegram. Commands express your desire in the quickest and most emphatic language possible.[59]

Finally, he told the officers that discipline and a well-developed sense of duty were inseparable and challenged the officers to develop their own sense of duty to the level exhibited by a thoroughbred horse. "You must develop the thoroughbred's sense of duty, otherwise you had better never have been born to wear the uniform you will inevitably disgrace." Duty, in Patton's view, was a privilege, and he told his officers they had better acquire a like frame of mind, for "to wear the uniform of an officer in the United States Army in any other frame of mind is to live a lie."[60]

The positive effect of Patton's exhortations is attested to by Pfc Melvin Winget, a driver and gunner who served in Company C, 345th Tank Battalion. Winget recalled that the level of discipline in his unit was exceptionally high and that "we had officers to be proud of." He described Patton as "gruf" but "behind us one hundred percent."[61]

Corporal Earl T. Carroll, who arrived at Bourg with the 328th Tank Battalion in September 1918, shared Winget's assessment of discipline in the AEF Tank Corps. He wrote later that when a soldier served under Patton or officers he trained, he "got it." Shortly after his arrival at Bourg, Carroll recalled, Patton told the men in his unit, "Why you God damned sons of bitches, do you think the Marines are tough? Well you just wait until I get through with you. Being tough will save lives."[62]

Patton's obsession with proper saluting also became legendary within the Tank Corps's ranks. Every tanker knew that if he had a chance to salute the colonel, Patton's return salute would be drill-manual perfect. It soon became a byword at Bourg that enlisted men encountering officers should "give 'em a George Patton [a sharp salute]."[63]

By May Patton was sufficiently satisfied with the operation of his tank center and school that he took the opportunity to go to the front with several of his officers to observe French tank operations. Early in the month he sent Brett and three other officers to the Montdidier sector for ten days. On the twenty-fourth Patton and seven officers visited the same sector, returning on 2 June.[64]

While at the front, Patton was able to meet with tankers from a French Schneider-equipped battalion who had supported the American 1st Infantry Division's attack at Cantigny. He learned the details of the battle from them and was impressed by the fact that not a single tank had been hit by German artillery fire. Col. Edward King, the 1st Division's chief of staff, discussed the operation with Patton from the Americans' perspective, and a Captain Johnson, who commanded the assault infantry working with the tanks, was "most enthusiastic" about the performance of the machines in battle.[65]

The French came through with an additional shipment of fifteen Renaults in May, allowing Patton to step up the pace of combat training.[66]

On 6 June, Patton had sufficient officers and men on hand to organize a second light tank battalion and create a larger staff. First Lt. Harry E. Gibbs became Patton's chief of staff, 1st Lt. Edmund N. Hebert the adjutant, 1st Lt. Maurice Knowles the reconnaissance officer, and 1st Lt. Robinson the supply officer. Patton turned over command of the 1st Light Tank Battalion to Captain Viner, who also served as the tank school's chief instructor. Company A was given to Capt. Ranulf Compton, Company B to Capt. Newell P. Weed, and Company C to Capt. Math L. English. Brett became commander of the 2d Light Tank Battalion, with Capts. Harry H. Semmes, William H. Williams, and Courtney H. Barnard taking command of Companies A, B, and C, respectively. Captain Ellis Baldwin was given command of the 301st Repair and Salvage Company, which serviced and repaired the center's tanks and wheeled vehicles.[67]

In June the tank school at Bourg implemented a nine-week program of instruction featuring thirteen courses. These included infantry drill; gas engines; tank driving; military intelligence; machine-gun; 37mm gun or six-pounder cannon; athletics and recreation; special advanced tank training; other specialty training; battle practice; inspections; fieldwork; and lectures.[68]

Infantry drill included "School of [the] Soldier"; squad, company, and battalion drill in both close and extended order; inspections and ceremonies; and marches and field training.

Gas engine and driver training included instruction on carburetor, magneto, clutch, transmission, engine, and differential theory and operation; conduct of minor repairs; and routine maintenance procedures. Tank foot drills and actual driving under increasingly difficult conditions were also included.

Military intelligence subjects included visibility of targets, range estimation, target designation, military topography, map reading and compass use, patrolling, messages, scouting, liaison work, signaling, and use of field telephones and pigeons.

During weapons training, the tankers learned the nomenclature and functioning of the 8mm Hotchkiss machine gun's various parts, how to assemble and disassemble the weapon, and how to perform maintenance and minor repairs. Live firing was conducted both dismounted and from tanks. A similar block of instruction was allocated for the 37mm gun and six-pounder cannon.

Ever conscious of the soldier's need for physical fitness, Patton saw to it that at least one hour per day was set aside for that purpose while the troops were in garrison.

During the seventh and eighth weeks, companies spent fifteen hours per week conducting simulated battlefield exercises covering all phases of tank training. An additional ten hours per week during the seventh and eighth weeks were allocated for the conduct of antigas drills, use of grenades and pistols, camouflage, and compass work.

Saturdays were devoted to inspections and ceremonies, marches, care of equipment, and personal and camp hygiene.

The highlight of the training program was a week-long "battle practice" conducted at the end. During this exercise the tankers conducted operations "under battle conditions . . . living under fighting conditions and making the preparation, participating in, and exploiting the success of a battle against a represented enemy."[69]

These "battle practices" gave the tankers a chance to maneuver their vehicles in tactical formations that Patton and other officers had worked out over a period of several months. Numerous trips to the front, coupled with lengthy discussions with British and French tank officers, provided Patton, Brett, and other Tank Corps officers with plenty of doctrinal food for thought. The British tactics centered on the heavy tank as a means, when employed in mass, to break through heavily defended positions, clearing wire obstacles and assaulting strong points and machine gun positions while the infantry followed closely behind. The French, on the other hand, chose to let their infantry lead the assault, relying on light tanks to come forward and reduce strong points and machine gun nests as necessary.

The tactics worked out by Patton and other members of the AEF Tank Corps staff incorporated ideas gained from both Allies. The basic scheme called for heavy tanks to lead the assault in the British manner, crushing wire obstacles, making paths across trenches, and reducing enemy strong points (see illustration). Each attacking heavy tank platoon was to be assigned a specific objective, to be broken down into subobjectives for each vehicle. Tank commanders were instructed to follow the creeping artillery barrage forward as closely as possible, seize their objectives, then wait under cover for light tanks or lead elements of infantry to join them before moving on to the next mission.[70]

Light tanks were assigned to follow about 100 meters behind the heavy tanks, in advance of the first line of skirmishers. Maintaining constant contact with each other, the light tankers and doughboys were to follow the heavies and "clean up everything that has been left behind by the large tanks." Contact between the light tanks and their supporting infantry was to be broken only if it became necessary for the tankers to advance and reduce any position hindering the infantry's forward progress. Additional light tanks were to be retained in reserve to assist the infantry unit's reserve element in exploiting any rupture of the enemy line.[71]

All of this looks simple on paper, and the tankers encountered little

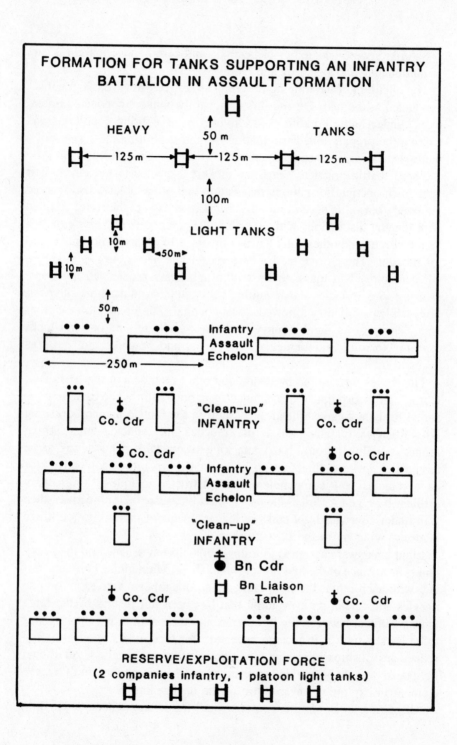

FORMATION FOR TANKS SUPPORTING AN INFANTRY BATTALION IN ASSAULT FORMATION

HEAVY   50 m   TANKS

125 m   125 m   125 m

100 m

LIGHT TANKS

10 m
50 m
10 m

50 m

Infantry
Assault
Echelon

250 m

Co. Cdr   "Clean-up"   Co. Cdr
INFANTRY

Co. Cdr   Co. Cdr

Infantry
Assault
Echelon

"Clean-up"
INFANTRY

Bn Cdr

Bn Liaison
Tank

Co. Cdr   Co. Cdr

RESERVE/EXPLOITATION FORCE
(2 companies infantry, 1 platoon light tanks)

difficulty executing the tactics in the training areas at Bourg. But Patton knew that his crews would have a difficult time when they went up against the Germans. Maintaining rigid formations in the face of a determined foe is an impossibility—a fact Patton acknowledged in a memorandum to his troops:

[Y]ou must ever strive to retain your line formation but you will never for one instant be able to do it. . . . [Y]ou must proceed like a polo team to do your job and then beat it as fast as possible to regain your formation which you will never do for when you get back in the line your neighbor on the right or left will be absent on a job of his own. Your sole guide then will be [the] direction of the distant objective some land mark you have previously picked out and which you must ever point towards in your periods of progression. What is asked of you is the most difficult of tasks— holding a formation which you never see except in your imagination. But it must be so indelibly stamped on your brain that gased and scared and tired and wounded as you will be it never for one instant escapes you. Your last conscientious effort must be to regain your formation and push on and ever on until there are no more Huns before you and the smiling vineyards of the RHINE open to your eyes.[72]

To ensure the tankers and infantry synchronized their efforts, Patton insisted that before beginning any operation the respective leaders meet and agree on the objective, the direction of movement, formations to be employed, and a plan for consolidation on the objective. Information on the activities of neighboring units also had to be exchanged, and points of contact for reports established. Patton also expected his officers to sound out the infantry leaders on the physical and morale status of their troops and determine a rate of attack consistent with the infantry's condition and the terrain to be crossed. Finally, tank and infantry commanders were required to coordinate the activities of machine gun battalions, artillery, and aviation units supporting the operation.[73]

By early June the tank companies at Bourg were using the twenty-five training tanks in rotation. The Renaults were prone to frequent breakdowns, which meant the mechanics in the tank center's repair and salvage company worked around the clock in order to keep as many as possible in working order. Despite their best efforts, rarely were enough operational to train more than two platoons at a time. Even under the best conditions, officers quickly became accustomed to beginning a maneuver with ten vehicles and completing it with only half still in running condition.[74]

Colonel Rockenbach's system for numbering units in the AEF Tank Corps was revised on 8 June, after word was received that the War Department had come up with a different scheme. This required renumbering the units in France. The 1st Tank Center thus became the 311th Tank Center, and the 1st and 2d Light Tank Battalions were redesignated the 326th and 327th Tank Battalions, respectively. [75]

At about this same time Brett moved his 327th Tank Battalion to Brennes, and Patton made arrangements for himself, Brett, Viner, and Gibbs to take the General Staff course at Langres.

Before departing, Patton initiated night-combat training. His tankers responded readily to the challenge. One company impressed him by completing a night march of more than six miles in less than three hours without deviating from the designated route. [76]

Rockenbach, who grew concerned when it became apparent in June that no American-built tanks would be available in quantity before 1919, prompted General Pershing to secure a promise from the French to equip Patton's two battalions with Renaults. It was the Tank Corps chief's intent to get at least one of his light battalions into combat when the AEF launched its planned offensive against the German St. Mihiel salient sometime in late August. [77]

This news did little to bolster the aggressive Patton's sagging morale. He again regretted that he had not sought an infantry assignment the previous fall. "The regiment I had a chance to join has been at it now for five months," he wrote in a letter to his wife. "Of course [I] have done a lot but I keep dreading lest the war should finish before I can realy do any fighting." Nevertheless, he still held some hope. He was certain that if the war would just last long enough for him to get into combat with his tanks, it would "work out greatly to my advantage[. B]ut the unknown is always full of terrors and I wake up at night in a sweat fearing that the d— show is over." [78]

The lack of tanks (and thus training opportunity) was also beginning to affect the morale of the men. Second Lt. Lester W. Atwood, at the time serving as first sergeant of Company A, 327th Tank Battalion, recalled that, after his unit moved to Brennes, "a parabola of the morale of the company, if such a thing could be depicted, would show somewhat of a declining curve during those summer months. Much fatigue [housekeeping-type duties] was the order of the day and this became more or less tiresome." [79]

Patton continued to attend to his duties as 311th Tank Center commander and commandant of the tank school during the months of July and August, while he attended the General Staff course.

In late July Patton drew a mild rebuke from Lt. Col. Daniel D. Pullen, Rockenbach's chief of staff, when he submitted a paper on tank tactics to GHQ AEF. Patton proposed that they abandon "the stereotyped formations heretofore thought essential." The Germans, he said, had repeatedly shown that it was possible to "break the enemy's initial crust of resistance" with a "sudden and violent" artillery barrage.[80]

Following this example, Patton said he thought tanks should be employed in conjunction with a bombardment similar to that used by the Germans but making "copious use of smoke" to cover their assault. The tanks should in turn be followed by infantry skirmishers, then infantry units in increasing density from front to rear. He recommended controlling the attack by progressively advancing the artillery barrage in increments of 200 to 400 yards or more, keeping up the fire on German defensive positions for thirty minutes to an hour, depending on the length of the jump. Because he envisioned the tanks operating for the most part in enemy territory, he suggested that heavy tanks be used to drag forward additional fuel supplies for the establishment of forward refueling points. Patton indicated in the report that he had already taken steps to conduct experimental training in the use of these tactics at his tank center.[81]

Pullen's response was succinct:

I do not think this is the time to propose any [new] tactics for the Tank Corps. Our first job is to get Tanks and then the second job is to get some Tank fighting units in the line. After we get some Tank units . . . in the fight we will be in position to talk about Tank tactics, and after we have been in one or more shows we will . . . be able to state exactly what we want and I believe what we say at that time will be accepted, while at the present time a great deal of what we say will be looked upon as hot air.[82]

The impulsive Patton had the wisdom to accept Pullen's advice in the spirit in which it was offered, and he turned his attention in the following weeks to subjects that would better prepare his men for battle using the accepted doctrine. During the first three weeks of August he composed and published at least four papers, including an update of his "Tank Drill Regulations (Provisional)," "Notes for the Guidance of Battalion and Company Reconnaissance Officers," "Tank Driving," and "Duties of the Platoon Leader."[83] Each of these papers dealt with the various subjects in minute detail and reflected Patton's own inimitable style.

By mid-August Patton had 900 men and fifty fully qualified officers at his tank center. They were all itching to get into the fight but still had only twenty-five tanks on hand. To harden the men physically and to help keep up morale, Patton began requiring each company to run a little over half a mile in columns of squads before breakfast each morning.[84]

As the long-awaited day when an independent American army would go into combat against the Germans approached, Patton could look back with satisfaction on the work he had accomplished. While Rockenbach commanded the AEF Tank Corps and advised Pershing and his staff on tank matters, Patton was the motivating force behind—and brains of—the light tank effort. Starting with nothing, he had created a highly trained, disciplined, and motivated light tank brigade. He and his officers had been required to design tables of organization and equipment for the units, recruit the men to fill them, develop tactics and doctrine for their battle-field employment, and devise a training program that would ensure they would be capable of functioning in combat. All this was done. That it had not been done on the scale originally envisioned was due to the inability of civilian industry to provide them with the equipment they needed, not because of any failure on their part.

Finally, on the morning of 20 August, Patton received the word he had been anxiously awaiting. A note was delivered to him during a lecture at the General Staff College directing that he report "at once to the Chief of the Tank Corps accompanied by your Reconnaissance officer and equipped for field service."[85]

It could mean only one thing—combat. More tanks must be on the way to Bourg, and with them the opportunity to lead his men into battle. Patton rushed to Bourg, flushed with excitement, where he appointed Viner assistant commandant and turned over command of the center.[86] Patton then set off for GHQ with First Lieutenant Knowles, his reconnaissance officer, to confer with Brigadier General Rockenbach, who had been promoted to that rank on 11 July.

Notes

1. George S. Patton, Jr., to Beatrice Ayer Patton, 23 Dec. 1917, as quoted in Blumenson, *Patton Papers,* 464.
2. Blumenson, *Patton Papers,* 464.
3. Col. George S. Patton, Jr., "History of Army Tank School, A.E.F.," 22 Nov. 1918, Modern Military Records Division, Record Group 120, Entry 22, Folder 229, NA, 1; ibid., 467.
4. Blumenson, *Patton Papers,* 469.
5. Patton, "History of Tank School," 2. The officers were 2d Lts. John D. Rice, Fred C. Winters, Robert T. Archer, Thomas D. Foley, Math L. English, Dan J. Sweeney, William H. Williams, Gus Struyk, Redmon, and Friend.
6. George S. Patton, Jr., to Beatrice Ayer Patton, 10 Jan. 1918, as quoted in Blumenson, *Patton Papers,* 470.
7. As quoted ibid., 471.
8. Ibid., 472.
9. Braine, "Personal Experience Report."
10. Patton, "History of Tank School," 2–3. The officers were 2d Lts. Edmund N. Hebert, Harry G. Borland, Courtney H. Barnard, Ellis Baldwin, Ernest A. Higgins, Henry F. Alderson, Robert J. Dunn, Robert C. Llewellyn, Theodore J. Sledge, Harry W. Bolan, and Brown.
11. Blumenson, *Patton Papers,* 473.
12. Patton, "History of Tank School," 3.
13. Blumenson, *Patton Papers,* 474–77.
14. George S. Patton, Jr., to Beatrice Ayer Patton, 25 Jan. 1918, as quoted ibid., 477.
15. George S. Patton, Jr., to Beatrice Ayer Patton, 14 Jan. 1918, as quoted ibid., 472.
16. Capt. George S. Patton, Jr., "Memorandum No. 1," 27 Jan. 1918, Patton Chronological Files, Box 9, 21–31 Jan. 1918, Patton Collection.
17. George S. Patton, Jr., to Beatrice Ayer Patton, 21 Jan. 1918, as quoted in Blumenson, *Patton Papers,* 475.
18. Patton diary entry dated 29 Jan. 1918, as quoted ibid., 479.
19. George S. Patton, Jr., to Beatrice Ayer Patton, 30 Jan. 1918, as quoted ibid.
20. Lt. Col. George S. Patton, Jr., "Provisional Light Tank Drill Regulations," 18 Feb. 1918, Modern Military Records Division, Record Group 120, Entry 1299, Box 25, 302d Tank Battalion Bulletin Board File, NA; Blumenson, *Patton Papers,* 481.
21. As quoted in Blumenson, *Patton Papers,* 485.
22. Ibid., 485–86.

23. Maj. George S. Patton, Jr., "Memorandum for the Adjutant," 23 Feb. 1918, Patton Chronological Files, Box 9, 18–24 Feb. 1918, Patton Collection.

24. Patton, "History of Tank School," 4.

25. 2d Lt. George B. Heilner, "Personal Experience Report," 14 Dec. 1918, Patton Military Papers, Box 47, Personal Experience Reports of Tank Operations– 1918, Patton Collection.

26. Patton, "History of Tank School," 4–5.

27. Patton, "Memo for Adjutant."

28. George S. Patton, Jr., to Beatrice Ayer Patton, 23 Feb. 1918, as quoted in Blumenson, *Patton Papers*, 491.

29. Maj. George S. Patton, Jr., "Memo to Col. S. D. Rockenbach," 2 Mar. 1918, Patton Chronological Files, Box 9, 1–15 Mar. 1918, Patton Collection.

30. "Memorandum for the Adjutant General," 5 Mar. 1918, Modern Military Records Division, Record Group 165, Entry 310, Box 216, File 7-61.13, NA.

31. Capt. Newell P. Weed, "Personal Experience Report," 11 Dec. 1918, Patton Military Papers, Box 47, Personal Experience Reports of Tank Operations– 1918, Patton Collection.

32. Memorandum, Maj. George S. Patton, Jr., to Maj. Daniel D. Pullen, 19 Mar. 1918, Patton Chronological Files, Box 9, 16–24 Mar. 1918, Patton Collection.

33. Memorandum, Maj. George S. Patton, Jr., to chief, Tank Corps, 16 Mar. 1918, Patton Chronological Files, Box 9, 16–24 Mar. 1918, Patton Collection.

34. Patton, "History of Tank School," 5.

35. Weed, "Personal Experience Report."

36. Blumenson, *Patton Papers*, 508–9.

37. George S. Patton, Jr., to Beatrice Ayer Patton, 24 Mar. 1918, as quoted ibid., 509.

38. Patton, "304th Brigade History," 5.

39. Blumenson, *Patton Papers*, 510.

40. Harry H. Semmes, *Portrait of Patton* (New York: Appleton-Century-Crofts, Inc., 1955), 39–40.

41. George S. Patton, Jr., to Beatrice Ayer Patton, 26 Mar. 1918, as quoted in Blumenson, *Patton Papers*, 510.

42. *Handbook of the Six-Ton Special Tractor Model 1917* (Washington, D.C.: U.S. Government Printing Office, 15 July 1918), 104–8.

43. Ibid., 14.

44. Memorandum, Maj. George S. Patton, Jr., to Col. Samuel D. Rockenbach, 22 Mar. 1918, Patton Chronological Files, Box 9, 16–24 Mar. 1918, Patton Collection.

45. Blumenson, *Patton Papers*, 511.

46. Patton, "304th Brigade History," 5.

47. Rockenbach, "Tank Corps Operations," 6.

48. General Order no. 5, General Tank Headquarters, AEF, 18 Apr. 1918, Modern Military Records Division, Record Group 165, Entry 310, Box 446, File 66-32.13, NA.

49. Blumenson, *Patton Papers,* 518.

50. Lt. Col. A.W. Bjornstad, "Memorandum," 20 Apr. 1918, Modern Military Records Division, Record Group 165, Entry 310, File 66-56.8, NA.

51. As quoted in Blumenson, *Patton Papers,* 524.

52. Patton, "304th Brigade History," 6.

53. Will G. Robinson to Lt. Col. Arthur J. Jacobsen, 6 July 1961, Patton Military Papers, Box 45, Miscellaneous Military Papers File, Patton Collection.

54. Ibid.

55. Blumenson, *Patton Papers,* 469.

56. As quoted ibid., 552.

57. Lt. Col. George S. Patton, Jr., lecture on "Discipline and Duty," undated, Patton Chronological Files, Box 10, 19–30 June 1918, Patton Collection.

58. Ibid.

59. Ibid.

60. Ibid.

61. Questionnaire, Pfc Melvin Winget, Company C, 345th Tank Battalion, World War I Survey Collection, Archives, USAMHI.

62. Questionnaire, Cpl. Earl T. Carroll, 304th Brigade, GHQ Troops, Company C, 328th Tank Battalion, ibid.

63. Semmes, *Portrait of Patton,* 42.

64. Patton, "304th Brigade History," 6.

65. Blumenson, *Patton Papers,* 537–38.

66. Patton, "History of Tank School," 6.

67. Blumenson, *Patton Papers,* 539; Rockenbach, "Tank Corps Operations," Appendix 3.

68. "Exhibit 'A' Schedule," undated, Patton Chronological Files, Box 10, 19–30 June 1918, Patton Collection.

69. Ibid.

70. G.H.Q. Tank Corps, "Notes on American Tanks," undated, Modern Military Records Division, Record Group 165, Entry 310, Box 217, File 7-61.5, NA.

71. Ibid.

72. Lt. Col. George S. Patton, Jr., "Points to be considered in the execution of a Tank Attack," 29 July 1918, Patton Chronological Files, Box 10, 22–31 July 1918, Patton Collection.

73. Maj. George S. Patton, Jr.(?), untitled paper on coordination of tank-infantry operations, Patton Chronological Files, Box 9, 18–21 April 1918, Patton Collection.

74. Semmes, *Portrait of Patton,* 40.

75. General Order no. 10, General Tank Headquarters, AEF, 8 June 1918, Modern Military Records Division, Record Group 165, Entry 310, Box 446, File 66-32.13, NA.
76. Blumenson, *Patton Papers,* 542.
77. Ibid.
78. George S. Patton, Jr., to Beatrice Ayer Patton, 13 June 1918, as quoted ibid., 543.
79. 2d Lt. Lester W. Atwood, "Personal Experience Narrative," 9 Dec. 1918, Patton Military Papers, Box 47, Personal Experience Reports of Tank Operations—1918, Patton Collection.
80. Lt. Col. George S. Patton, Jr., untitled paper dated 22 July 1918, Patton Chronological Files, Box 10, 22-31 July 1918, Patton Collection.
81. Ibid.
82. Lt. Col. Daniel D. Pullen to Lt. Col. George S. Patton, Jr., 25 July 1918, Patton Chronological Files, Box 10, 22-31 July 1918, Patton Collection.
83. See Patton Chronological Files, Box 10, 29-31 July 1918 and undated, Patton Collection.
84. Blumenson, *Patton Papers,* 562.
85. As quoted ibid.
86. Ibid.

CHAPTER III

The Heavy Tanks:
A Gathering at Wool

While Patton was busily preparing to open the Light Tank School at Langres in early 1918, concurrent efforts were under way to recruit and train officers and men to man heavy tanks. Because the British were the only allies employing heavy tanks, it was decided that the British Home Training Center at Wool, England, was the logical place to train the initial contingent of American heavy tankers for the AEF, and Rockenbach obtained the approval of the British and the AEF staff to establish an American training center as an annex to the British Tank School at neighboring Wareham.[1]

Rockenbach cabled the War Department in early February with a request that fifty officers and 100 NCOs be selected and sent to England as soon as possible so they could be trained to serve as the cadre for a force he hoped would grow to five heavy tank battalions by July. He requested that the men to fill those battalions be recruited from units in the United States and that the first battalion be shipped to England by 1 March, with the remainder to follow at a rate of one per month.[2]

Acting on Rockenbach's request, the Army activated the 65th Engineer Regiment at Camp Meade, Maryland, and issued a call for "high-quality" volunteers for service with tanks. The type of men sought for duty in what some called "President Wilson's slaughterhouses" were those of a "daring and adventurous spirit, . . . unafraid in any dilemma," and who were "cool and calculating and willing to take the long chance."[3]

Soldiers from all branches of the Army responded enthusiastically to the call. As many as 250 signed up in a single day at Camp Meade, and officers at Camp Devens, Massachusetts, were deluged with applications. A single afternoon's roster of recruits at one New York headquarters included

> two gold miners, three Boer War veterans, three professional pugilists, six members of last year's varsity football team at Williams College, six former United States Marines, three men who had won the French Croix de Guerre, a filibuster, an Argentine cavalryman, a dancing master, a lion tamer, and forty men from the University of Chicago who had enlisted as an ambulance unit and then transferred.[4]

By the end of February, enough men had been recruited to organize Companies A, B, and C of the 1st Separate Battalion, Heavy Tank Service, 65th Engineers, and D Company, 2d Battalion, Heavy Tank Service, 65th Engineers.[5]

David A. Pyle was typical of those who volunteered. Pyle, then serving as a corporal with D Battery, 310th Field Artillery, was awaiting orders to attend Officer Training School when he heard that volunteers were being sought for duty with heavy tanks. He recalled having "considerable trouble" getting an interview at Camp Meade but was finally accepted and assigned to A Company of the 1st Separate Battalion as company clerk.[6]

The tankers at Camp Meade were kept busy with drill, physical training, hikes, and inspections while awaiting shipment to England.[7] On 5 March 1918 they were detached from the Corps of Engineers and assigned directly to the Tank Service, concurrent with Lieutenant Colonel Welborn's appointment as director of the Tank Service in the United States.[8]

In France, meanwhile, Rockenbach selected Lt. Col. Conrad S. Babcock to command the 2d Tank Center, to be established at Bovington Camp near Wool.[9] Patton and Rockenbach then departed for Wool on 5 March, where they spent a week conferring with the British, observing training at the British Tank School, and preparing for the arrival of the first heavy tank battalion from the United States.[10]

One of the officers who had been called to assist with the 1st Battalion's formation at Camp Meade was Capt. Dwight D. Eisenhower. In mid-March, when word arrived that the battalion should begin preparation for overseas movement, Eisenhower was sent to New York to coordinate with port authorities. After two days of hard work attending to the details involved with embarkation and shipment, Eisenhower returned to Meade, where he found, much to his dismay, that he would not be going with the battalion.

"My chief said he was impressed by my 'organizational ability.' I was directed to take the remnants of the troops who would not be going overseas, and proceed to an old, abandoned campsite in Gettysburg, Pennsylvania, of all places."[11] The men who accompanied Eisenhower were in D Company, 2d Battalion, which became the nucleus of the 302d Tank Battalion at Camp Colt.

The tankers of the 1st Battalion got their marching orders on 23 March and moved by train to Camp Merritt, New Jersey, then on to Hoboken, where they were ferried to New York City on the night of the twenty-sixth. On the twenty-seventh they boarded the White Star liner *Olympic,* already crowded with 7,500 troops bound for France. [12]

As the clerk for A Company, Corporal Pyle found his accommodations to be considerably better than those of his fellow soldiers. He joined his company executive officer, Second Lieutenant Hilliard, in the stateroom belonging to 1st Lt. John M. Franklin and was told they would "set up shop here" and that Pyle would be quartered with them for the voyage. He found out later the reason they had been provided such luxurious living quarters was that Franklin's father was president of the White Star Line. "I made the trip living high on the hog and was more or less ashamed of my good luck."[13]

The *Olympic* made an uneventful crossing, arriving in Brest, France, shortly after dawn on 7 April. The day was spent unloading equipment and all of the troops except the men of the 1st Battalion. That night the *Olympic* slipped out of Brest under cover of darkness and headed for Southampton, England, arriving at dawn off Plymouth. "[We] had a submarine scare which called us all to stations, but after [much] destroyer activity and commotion we got our first view of the White Cliffs and Plymouth Harbor."[14]

At high tide, the *Olympic* steamed up the narrow channel to Southampton, where the men disembarked and formed up under a shed beside the London and Southampton Railroad tracks. Here they had their first encounter with a British soldier:

A Scottish Sergeant Major came strutting down the platform with his colorful kilts and his swagger stick under his arm. When our men saw him they began to direct many remarks such as "She's going to sleep with me tonight," or "He's going to be my girl tonight." Of course this was enough to insult the King and Queen, so he dashed back and demanded to know who was the bloody bugger who said that. About a dozen men broke rank[s] and the last I saw of him he was running off the end of the platform into the streets of Southampton with several officers trying to restore order and get the company back into line.[15]

The next day, 9 April, the battalion was taken by train to Wareham. During the three-mile march from the station to their new camp, the tankers were greeted warmly by the townspeople, and, as they passed the village school, the headmaster led his students in "three hearty cheers" while one youngster proudly waved an upside-down American flag, oblivious to the fact that this was a signal for distress.[16]

The first week at Wareham was devoted primarily to orientation training. The men of the 1st Battalion learned the layout of the British Tank School and were told what they could expect during their training—"a very comprehensive program" involving all phases of tank operations from the theoretical to practical application of mechanical maintenance, tank driving and gunnery, reconnaissance, map and compass reading, gas warfare, signaling techniques, battle drills, and so on. They also received their first cash payment in British currency. While the Americans initially found the English monetary system perplexing, their gambling activities quickly taught them the value of the British pound.[17]

Two things in particular caught the Americans' attention during their first week at Wareham: the presence of women from the British Women's Auxiliary Corps at the camp, and the bedding the men were provided, which consisted of three long boards supported by two small trestles on which lay a simple, straw-filled mattress.[18]

Training for the 1st Battalion's tankers began during their second week at Wareham. Conducted by white-sweatered instructors from the British Tank School, it consisted mostly of bayonet drill, semaphore signaling, and "physical torture"—fitness training administered in "huge doses."[19]

This increased pace made the Americans painfully aware of the "woefully inadequate" British diet. Accustomed to hearty garrison rations back in the United States, they found "the regime of slum [a concoction the contents of which are not described], tea, cheese, bread, and occasionally as a great luxury, jam" to be far from filling. Even worse was the English mutton. The tankers, used to a steady diet of American beef, speculated that it was actually goat meat. In response to their heated complaints, the

British maintained their menu was the same as "Tommy's." Nevertheless, food proved to be a continual source of irritation, and many men bought bread, eating it plain to satisfy their hunger. Others elected to raid the kitchen late at night in an effort to ease their pangs.[20]

On 18 April, in response to Rockenbach's directive on Tank Corps unit designations, the 1st Battalion became the 41st Heavy Tank Battalion.[21]

Pyle and a number of other officers and NCOs were selected early in the course to attend specialized instructor training, the object of which was to prepare them to establish a similar tank school at Camp Colt. Although this school was organized, it later proved necessary to retain those who graduated from the instructor course in England for duty at Bovington, where they trained both British tankers and Americans of the 303d Tank Battalion, which had been organized at Camp Colt in May and arrived at Wareham in August.[22]

By 6 May, the first group of American tankers was ready to move on to more advanced training at Bovington. There they received classes on tank driving, compass use, employment of pigeons, camouflage, minor repairs, machine guns, the six-pounder cannon, gas warfare, and mechanical training. Models showing the inner workings of the various automotive mechanisms were used along with plans and charts during the mechanical instruction.[23]

The highlight of the training at Bovington was tank driving. The men were willing to endure long, hard days of marching and maintenance while waiting for their turn to take the controls of one of the twenty-six-foot Mark IVs. Two men, one on each side of the tank to operate the gears, were required in addition to the driver to make the armored behemoth move. Few things they had experienced could compare with

> the thrill of pulling a monster out of a deep trench, nose pointing to the sky with the engine's deafening roar, the acrid, never-to-be-forgotten smell of exploded gas, scorching oil and grease, and hot steel, the quick shutting down of the throttle, the gentle swing to earth, and then the triumphant roaring answer of the engine to the opening of the throttle and the more-the-merry clanking of the track plates.[24]

To further add to the realism of the training, their British instructors replicated the battlefield in "minute detail." Little was left to the imagination. "Shell holes" were blasted with mines, trenches identical to those they would encounter in France were dug, and other obstacles such as barbed wire were emplaced. Over these the tankers drilled and drilled again.[25]

While this first group of tankers was tackling the advanced training at Bovington, Patton received a letter from Capt. Ralph I. Sasse, a cavalry officer in the 1st Infantry Division, who sought duty with tanks. Sasse explained that he had made four applications for transfer but had been denied each time because he "was needed here." He pleaded with Patton to provide any assistance he could to secure the "successful approval of my transfer." Patton was able to obtain Sasse's transfer to the Tank Corps, and Sasse would go on to command the only American heavy tank battalion to see combat during the war.[26]

The 41st Heavy Tank Battalion became the 301st Tank Battalion on 8 June, when the AEF Tank Corps was forced to change its unit designation system to bring it in line with the system inaugurated by the War Department in the United States. The 2d Tank Center also changed its designation, becoming the 301st Tank Center.[27]

Later in June, when it became apparent that delays in the production of Liberty aircraft engines would postpone production of the first joint American-British Mark VIII heavy tanks in France until at least 1 October,[28] Rockenbach coordinated with the British for tanks with which to equip the 301st Tank Battalion and others soon expected to arrive in England for training. The British, experiencing their own shortages of heavy tanks, agreed to provide only enough to equip the 301st, and then only if the battalion were attached to the British Expeditionary Forces when it deployed to France.[29]

During more than two months of training under the watchful eyes of their allies, the Americans had made an impression on the British. It was not a good one. Pyle recalled that the British "called us Uncle Sam's Rag Time Army, which pretty well presents my very own view." It was plainly evident that the high standards of discipline enforced by Patton at the Light Tank School in France were not required by his counterparts in England. This may account for why Pyle says the men of the 301st held their commander, Lt. Col. Henry E. Mitchell, in such low esteem—although he says they considered Major Sasse and their other officers to be excellent.[30]

In early July, after the British agreed to equip the 301st Tank Battalion, driver training was conducted on the more advanced Mark V tank at Sanford, about a mile east of Wareham. The Mark V was easier to handle than the Mark IV, requiring only one man to operate it. The level of gunnery training was also increased as tankers rotated through firing ranges at neighboring Hyde Heath and conducted live-fire battle runs from moving tanks at nearby Lulworth.[31]

The AEF staff, which learned from the experience of the 1st and 2d Infantry Divisions at Soissons in mid-July that tanks would play an important role in future operations, authorized the Tank Corps to increase its paper strength. The new organization for the Tank Corps, approved later that month, called for an additional five heavy tank battalions. These, added to the five previously authorized heavy battalions and twenty light battalions, would be incorporated into ten tank brigades, each consisting of a separate headquarters, one heavy battalion, two light battalions, and a repair and salvage company.

Tank Corps troops were considered GHQ troops and were to be allocated to the field armies based on terrain and the nature of planned operations. An army's "slice" would consist of an Army Tank Corps headquarters, a tank training center, a heavy artillery ordnance mobile repair shop, and five tank brigades.[32]

Advance elements of the 303d Tank Battalion arrived in early August, as the men of the 301st completed their training program and began to prepare for deployment to France. Rockenbach conducted an inspection tour of the 301st Tank Center on 8 August, closely followed by Brigadier Glasgow, commander of the British Tank Center. Both found all to be in order.[33]

During the early weeks of August, as the 301st awaited movement orders, several wags, amused by Patton's efforts to make the tankers' appearance distinct from that of other soldiers in the AEF, developed their own idea of an "official uniform—Tank Corps." They envisioned an overseas cap with a green tassel dangling over the left eye; a shirt cut along the "same lines as those worn by [the] Chinese Labor Corps," with elbow-length sleeves and "Prince Albert swallowtails" and hasps and track pins in lieu of buttons; trousers made from olive drab tarpaulin, cut obliquely halfway between the knees and ankles, worn high enough "to slip arms through and drape from shoulders," and with a red stripe down the left leg and a yellow stripe down the right; pale blue field shoes featuring three-eighth-inch bolts instead of hobnails and track plates instead of heels, to be worn without socks; decorative devices including "colors on shoulders and diamond between shoulder blades," a six-by-eight-inch board to be suspended over the chest from the neck for the wear of medals, chevrons "worn on rear of left pants leg amidships," with corporal chevrons to be worn under an inverted teacup and sergeant chevrons under clasped hands; authorization for privates only to carry canes; and wearing of the regulation pistol belt but with every hole having "suspended from it useful articles, such as cork screws, can openers, button hooks, etc." The tank officer's uniform would be the same, except for decorative devices. Officers, most of whom had been commissioned from the enlisted ranks, would "wear a model Christmas tree suspended from their bar or bars showing the origin of their rank."[34]

Off-duty activities included trips to London, Dorchester, and Bournemouth. Some of the tankers, including Pyle, dated local women, "like that saucy little Margarete Watkins, daughter of the Governor of Lloyd's bank in Wareham." Two of the companies based at Wareham and Bovington boasted "a number of talented actors and one producer from the American Theater." They put on a musical comedy in Bournemouth that drew a capacity crowd, and "Del Puienni . . . brought down the house" with his rendition of "On the Road to Mandalay."[35]

The AEF Tank Corps's situation by 20 August, when Patton got the word to report to Rockenbach at Chaumont to begin preparation for the St. Mihiel offensive, was far from ideal. Although authorized thirty battalions on paper, the Tank Corps had only three—the 326th and 327th Light Tank Battalions and the 301st Heavy Tank Battalion—trained and ready for action. Worse yet, none of these units owned a single fighting tank. No additional light tank battalions had yet arrived in France, although four were en route from the United States, and only one other heavy tank battalion, the 303d, was in England.[36]

On 23 August the 301st embarked at Southampton and sailed for Le Havre, arriving early the next morning. At Le Havre the tankers boarded "forty-or-eights," railcars designed to carry either forty men or eight horses, bound for the British tank center at Bernecourt. "Gee," said one tanker as he looked around his crowded car, "I wish I was a horse."[37]

While the battalion was at Bernecourt, several officers and NCOs were sent to the front near Arras to gain experience with a Canadian tank unit. Two others, Sergeants Hiebard and Abbot of Company C, were killed by shellfire while advancing with British tanks at 2 A.M. on 1 September in the vicinity of Wancourt.[38]

First Lt. Carleton Reynell, who had been a mechanical instructor at the 301st Tank Center, attended the British reconnaissance course shortly before the battalion deployed to France. Upon arrival at Bernecourt he was relieved of his duties as the 301st Battalion reconnaissance officer and attached to the British 15th Tank Battalion. During his stay with the 15th, Reynell participated in three attacks against the "strongly fortified" town of Vaulk-Vrancourt, about three miles northeast of Bapaume. Working with the 15th's reconnaissance officer and deputy commander, Reynell watched the first two attacks reach their objectives only to have the war-weary British infantry driven back by violent German counterattacks. At one point a British general told him, "I wish we had American or Australian infantry here to hold this place. Our men are too played out."[39]

The remainder of the battalion moved to Erin on 30 August, where the crews were issued Mark V tanks by the British, then spent a week cleaning, oiling, and making minor repairs on the vehicles before entraining for Bihucourt on 6 September. They detrained the next morning at Achiet-le-Grand, about two miles east of Bapaume, and moved by road to Bihucourt. There the troops found a town that had been destroyed by shelling, so they had to be billeted in damp, muddy, rat-infested dugouts left behind by New Zealand soldiers who had moved up to the front. There were no washing facilities or furnishings, so the tankers used gasoline cans for seats, wash pans, pillows, bedposts, cupboards, shelves, tables, and shaving mugs.[40]

The battalion continued to perform maintenance on tanks and conduct classes while waiting for orders that would send them to the front. They were visited by several British officers, who shared their experiences with the men. One of them described what it was like to be in a tank being hit by machine-gun fire. He said that as the bullets hit the hull, thin flakes of steel, called "splash," flew off inside the tank, getting into the faces and hands of the crew, stinging them. To soothe the pain and prevent infection, the crewmen were required to bathe the wounds with iodine. In addition, "simultaneously with the sound of impact and creation of the flakes, a blue flame is seen, and when a machine gun plays up and down the joints" in search of a weak seam, the tank's interior "looks like a demonstration of cheap fireworks." It was hardly a thrilling prospect.[41]

The experiences of the heavy tankers as they trained for combat differed greatly from those of their light tank counterparts in France. The critical difference was that they were situated far from other AEF units and conducted the majority of their training under the watchful eyes of the British, frequently with British instructors, and always on British equipment. Because the decision was made to employ them in combat under British control, there was little incentive for American officers in England to take as analytical an approach to tactics and doctrine or training as Patton was forced to take in France. Their whole program was practically spoon-fed to them.

The danger in such a situation is that complacency may so overcome an officer's judgment that he may overlook obvious defects in theory or practice, especially when dealing with a weapons system still in its infancy. Fortunately, this was not the case with the heavy tank officers. They continually questioned the British and sought to adapt their ideas to American organization and tactical doctrine should the war last long enough for them to get into combat under the control of an American army commander. To this end, a steady dialogue was maintained via correspondence and visits to Rockenbach's headquarters.

In any case, there was little disagreement over the tactical employment of heavy tanks. Their mission was simple: to lead the infantry into the trenches and clear out barbed-wire obstacles and enemy machine-gun nests as they were encountered. This was to be accomplished by attaching a single tank platoon to each assault infantry battalion in the first wave.

As September 1918 passed its midpoint, word filtered down to the men of the 301st that they would be joining the British Fourth Army in an attempt to breach the Hindenburg Line in the vicinity of St. Quentin. This was welcome news, as they had been anxiously watching the exploits of their light tank brethren in the St. Mihiel offensive and were champing at the bit to get into the fight.

On 21 September the battalion again moved to Achiet-le-Grand and loaded onto trains for shipment to a new camp in the vicinity of Manancourt. Once there, the 301st's tankers continued to prepare themselves and their machines for the ordeal that lay ahead.[42]

Notes

1. Rockenbach, "Tank Corps Operations," 5.
2. Ibid., 6; Rockenbach, "The Tank Corps," 6.
3. *A Company, 301st Tank Battalion History* (Philadelphia: E. A. Wright Co., 1919), 188–89.
4. Ibid., 189.
5. Ibid., 32–34; *Order of Battle of the United States Land Forces in the World War (1917–19), Zone of the Interior,* Vol. 3, Part 2 (Washington, D.C.: U.S. Government Printing Office, 1949), 1543–44.
6. Oral History, Sgt. David A. Pyle, 301st Tank Battalion, World War I Survey Collection, Archives, USAMHI.
7. *301st Battalion History,* 34.
8. "Memorandum for the Adjutant General," 5 Mar. 1918, Modern Military Records Division, Record Group 165, Entry 310, Box 216, File 7-61.13, NA.
9. General Order no. 3, General Tank Headquarters AEF, 2 Mar. 1918, Modern Military Records Division, Record Group 165, Entry 310, Box 446, File 66-32.13, NA.
10. Patton, "304th Brigade History," 5.
11. Dwight D. Eisenhower, *At Ease: Stories I Tell to Friends* (Garden City, N.Y.: Doubleday & Co., Inc., 1967), 137; *Order of Battle,* 1544.
12. *301st Battalion History,* 34–35; Pyle, Oral History.
13. Pyle, Oral History; Questionnaire, Sgt. David A. Pyle, 301st Tank Battalion, World War I Survey Collection, Archives, USAMHI.
14. Ibid.
15. Ibid.
16. Ibid.; *301st Battalion History,* 42.
17. Pyle, Oral History.
18. *301st Battalion History,* 42.
19. Ibid.
20. Ibid., 42–43.
21. General Order no. 5, General Tank Headquarters AEF, 18 Apr. 1918, Modern Military Records Division, Record Group 165, Entry 310, Box 446, File 66-32.13, NA.
22. Pyle, Oral History; *Order of Battle,* 1545; Capt. Carleton Reynell, "Personal Experience," 13 Dec. 1918, Patton Military Papers, Box 47, Personal Experience Reports of Tank Operations–1918, Patton Collection.
23. *301st Battalion History,* 46.
24. Ibid.
25. Ibid., 195.

26. Capt. Ralph I. Sasse to Lt. Col. George S. Patton, Jr., undated, as quoted in Blumenson, *Patton Papers,* 531.

27. General Order no. 10, General Tank Headquarters AEF, 8 June 1918, Modern Military Records Division, Record Group 165, Entry 310, Box 446, File 66-32.13, NA.

28. Cable no. 1636-R, 29 June 1918, attached as enclosure to Collins, "Development of the Tank Corps."

29. Rockenbach, "Tank Corps Operations," 7.

30. Pyle, Questionnaire.

31. *301st Battalion History,* 46–47.

32. Rockenbach, "Tank Corps Operations," 6–7.

33. *301st Battalion History,* 47.

34. Ibid., 138.

35. Pyle, Oral History.

36. Rockenbach, "Tank Corps Operations," 7.

37. *301st Battalion History,* 50.

38. Ibid.

39. Reynell, "Personal Experience."

40. Ibid., 50–51.

41. Ibid., 197.

42. Ibid., 54.

CHAPTER IV

Training in the United States: A Tank Corps without Tanks

C apt. Dwight D. Eisenhower, whose dreams of battlefield glory were dashed when he received orders to create the first Tank Corps training center in the United States in early March of 1918, recalled that his "mood was black" as he made the journey from Camp Meade to Gettysburg, Pennsylvania, with his small detachment left over after the departure of the 1st Battalion, Heavy Tank Service, 65th Engineers, for England.[1]

He would not have been quite so despondent, however, had he known what the future held in store for him. Although he commanded only a handful of men, had no equipment, and had been assigned an unused plot of land on the site of Confederate Maj. Gen. George E. Pickett's famous charge at the Battle of Gettysburg, within just seven months he would be a lieutenant colonel commanding a training post with more than 10,000 enlisted men and 600 officers. The experience he gained in this assignment would prove far more valuable to him in his later career than any battlefield exploits he might have accomplished as an infantry company or battalion commander in France.

When Eisenhower and his tankers arrived at Camp Colt, one of their first acts was to hoist the flag on an old pole that someone had left there. Watching the ceremony, Eisenhower noticed a Captain Garner, a former senior noncommissioned officer who had been promoted from the ranks a year earlier, quietly standing off to the side. As he approached Garner, whom he did not know well,

> he said, without looking at me, "Captain, the last time I was on this ground was many years ago. At that time I was standing before a general court-martial which sentenced me to six months in the guardhouse, and then suspended the sentence.
> "Now," he said, "I'm a captain in that same Army, and I'm standing here [in the same] camp in which I was disgraced." As I looked up, this old, hard-bitten, gray-headed former noncom had tears streaming down his face.[2]

Garner epitomized the type of officer with whom Eisenhower would be working during the coming months. All of them were volunteers for the fledgling Tank Corps, and many came from the enlisted ranks, bringing with them the necessary knowledge of infantry and close-order drill, as well as army life, needed to mold into soldiers the thousands of young men who would soon be pouring into the new post.

The first task Eisenhower and his men attended to was the creation of living and working quarters. This was speeded by the ready availability of tentage. Although they had no building materials with which to make tent frames or platforms, they soon had a functioning camp ready to handle the expected influx of volunteers, whom Eisenhower was instructed to equip, organize, and train. Adding to the difficulty of the task was the fact that Camp Colt was considered both a mobilization and an embarkation point, meaning it had the added administrative work load of processing men for direct overseas shipment.[3]

By April, Eisenhower had more than 500 men at Colt, including elements of the 302d Heavy Tank Battalion and the 326th, 327th, and 328th Light Tank Battalions, all of which he expected would be ordered overseas at any time. Unfortunately, the lack of tanks available in France, coupled with the Allies' decision to give transport priority to infantry and machine-gun units, meant the tank troops would stay at Colt indefinitely.[4]

A sudden snowstorm hit the ill-prepared post in early April. None of the tents had stoves, and Eisenhower was forced to make his way through the drifts into town, where he proceeded to buy up every stove available. The trip, which normally took just five minutes by automobile, required more than two hours. There still were not enough stoves to go around, so Eisenhower instructed the men to build stone fire pits in the remaining tents, using rock and shale from the piles built up when they had cleared the area. "These permitted modest fires inside the rest of the tents, although they required a patrol of men all night to keep the fires going and make sure no tent burned down."[5]

Although he expected an outbreak of illness from the bitter cold, such was not the case. Nor were there many complaints about the Army's "negligence" for exposing the men to "premature death." Instead, the men "seemed to take the storm as a splendid way to demonstrate their robust health" and were soon "ready to tackle anything in the way of work."[6]

But aside from basic drill and housekeeping chores, there just was not much to do. There were no tanks or weapons on hand for training, and the Tank Corps in the United States was not in contact with the AEF Tank Corps, so there was no guidance available on how best to prepare the men for tank combat. It was, in effect, a Tank Corps in name only. Each of the Corps's training centers provided little more than basic soldier training. Colonel Welborn's headquarters in Washington was concerned primarily with recruiting and production and served as the coordinating point for stateside training activities. It was understood by all concerned that tanks would be shipped to France as fast as they came off assembly lines and that actual combat training would have to wait until units deployed to France.

Faced with the dilemma of keeping morale up and preventing the "dry rot of tedious idleness," Eisenhower and his officers turned to newspapers and magazines for ideas and used their imaginations to devise courses of instruction. From these brainstorming sessions emerged a telegraph school. A shortage of skilled Morse Code–trained operators led Eisenhower to believe that a "first-rate telegrapher might prove himself worth more than a company of riflemen." A motor school was also organized. Experienced mechanics culled from the ranks provided instruction using "secondhand motors of all kinds" obtained from the local community as training aids.[7]

During one of his twice-weekly visits to Tank Corps headquarters in Washington, Eisenhower was able to obtain several swivel-type Navy "three-pounder" guns with which the men trained despite the lack of ammunition. A shipment of machine guns arrived shortly thereafter, and soon men were able to assemble and reassemble them blindfolded. Machine-gun ammunition was obtained, and then

> someone had the notion of mounting the machine-guns on truck trailers
> or on flatbed trucks, and so we were able to train the men to fire from

mobile platforms at both moving and still targets. The only satisfactory place for firing was Big Round Top. . . . Its base made a perfect backstop. Soon, soldiers were shooting from moving trucks at all kinds of targets there.[8]

Troops continued to pour into Camp Colt, and by the end of May the 303d and 304th Heavy Tank Battalions and the 329th and 330th Light Tank Battalions had been added to the rolls.[9]

A camp newspaper was created in May. Called *Treat 'Em Rough,* it included items about camp activities as well as reprints of motivational articles appearing in the civilian news media. One such article, originally published in *Every Week,* was headlined "Picking the Right Men for the Tanks." The author described the men sought for tank duty as those who "will fight bravely and intelligently, where other men would give up." He suggested that recruiters size up potential tankers critically:

> Height matters little: he may be anywhere from five to six feet tall. But his eyes should be gray-blue. If they have a steely or glassy cast, so much the better. And he should "smile with his eyes" when you talk danger to him. He should be a man whose women-folk believe in his fighting qualities and do not hold him back. If he has a good record in sports, big-game hunting, or anything else that requires backbone and stamina, better still.[10]

Shortly after this story appeared in print, Eisenhower added his own thoughts about what a tanker should be. He agreed that tankers must be brave but debunked the idea that hair or eye color indicated a predisposition for bravery. Instead, said Eisenhower:

> A brave man can be spotted at once by his carriage. "Get the carriage of your men" is my first order to the officers. The carriage is stomach in, chests out.
> A brave man is one who faces danger, dreads it, but faces it because he believes there is something bigger in life than passing out of it, no matter how painful the passing may be. He is a fighter. An upstanding initiative, forceful personality with self-confidence.[11]

This interest in soldierly bearing was manifested frequently in Eisenhower's lectures to his officers and men. In a memorandum he ordered read by company commanders to each company on 20 June 1918, Eisenhower chided the troops for "showing a lack of the snap and precision which must characterize close-order drill." He attributed the problem to poor instruction and directed leaders at all levels to "strive by voice, example, and untiring energy to instill the soldierly spirit and pride in the men, which is the true secret of good close-order drill."

Pressures brought on by the Tank Corps's high visibility in the civilian news media and in military circles also concerned Eisenhower. He wanted soldiers to "be reminded constantly that they belong to the *Tank Corps* [emphasis in original]" and that as "a new and special branch of our Army" they would be "closely watched." As members of "the foremost fighting units of all fighting troops," they could expect close scrutiny by members of the other branches "to see what manner of man is entrusted with this most important mission."

Colonel Welborn had earlier indicated that the Corps was expected to become self-sustaining by supplying the majority of its officers from the enlisted ranks. Eisenhower used this in an effort to motivate his men, noting that the soldier "who shows in his appearance that he has the proper pride in himself, the Tank Corps, and the Army, has a big start toward a commission." [12]

Despite this emphasis on strict discipline, recalled Sgt. Maj. Claude J. Harris, the senior enlisted man at Camp Colt, the soldiers responded positively to Eisenhower's concerned leadership. Harris noted that the young commander's decisions "affecting the welfare of his officers and men were always well tempered." Harris said Eisenhower was also able to quickly estimate a man's potential and find an appropriate place for him in the organization. In the few instances where Eisenhower's judgment proved wrong, he called the man in, pointed out his errors, and made adjustments "to suit the situation." This tactful approach, according to Harris, earned for Eisenhower the admiration and loyalty of his officers. [13]

Eisenhower was distressed by word that German propagandists were alarming civilians by spreading "tales of hardships in the camps of poor food, of disease, and sometimes [going] so far as to call up anonymously on the telephone to tell the relatives of the soldiers that their lads are dead." This was especially distressing when considered in light of the "surprisingly small amount of written news or comment" communicated by his men to their relatives at home. The few letters that did go out were usually dull and told little about what training and camp life were really like.

To combat this problem, Eisenhower drew on the experience of units overseas. He suggested that company commanders select editors to produce a weekly "company letter" based on contributions from all of the soldiers in the unit. These would then be compiled at camp headquarters and sufficient copies made to provide several for each man. The soldiers were then encouraged to add postscripts to make the printed letter more personal.[14]

The troops' off-duty conduct in neighboring civilian communities was a continuing source of irritation. Complaints were received from residents of Hanover, York, Gettysburg, and Chambersburg. Soldiers were accused of showing "utter disregard" for civilian sensibilities, parading drunkenly and cursing loudly in the streets, as well as "accosting ladies." Such actions were, in Eisenhower's view, attributable to "thoughtlessness and inexperience rather than to viciousness," and he believed disciplinary action could be avoided if commanders simply apprised members of their units of "the bad taste and vulgarity of such actions."[15]

Not all of the friction was caused by soldiers, however. Although most area residents welcomed the men from Camp Colt, "if for no other reason than patriotism," some "saw in them a source of quick and easy profit, catering to all their appetites."[16]

One of Eisenhower's first clashes with local businessmen stemmed from a War Department directive authorizing camp commanders to place off limits saloons and other establishments serving liquor within a five-mile radius of the installation.

Rather than take this step, which military police officials thought would cause a "surge in bootlegging, a business they couldn't control," Eisenhower elected to allow tavern owners to police themselves, asking them not to sell intoxicants to men in uniform. He warned them that failure to comply would result in his application of more stringent measures.

While most acceded to his wishes, one hotel owner continued to "surreptitiously" sell liquor to off-duty troops. Unsure of his authority in the matter because the establishment was a hotel rather than a saloon, Eisenhower posted a provost guard around the building and ordered them to allow no soldiers to enter it. This had an immediate impact, as it meant visiting "brothers and husbands and sons and sweethearts would not go into the hotel either." The owner, seeking relief, met with Eisenhower and promised to comply with the edict if given another chance. Eisenhower relented, but the owner failed to uphold his end of the bargain.

Again the provost guard was mounted, and this time Eisenhower refused to withdraw it. Angered, the innkeeper sought the aid of his congressman. At their next meeting, which included the congressman and his secretary:

> I listened [to the innkeeper's] story and said that I stayed with my position. After debate, the congressman said, "Well, we have means—we can go to the War Department. If you're going to be so stubborn, I'll have to take up the question of replacing you."
> I said, "You do just exactly that."
> He looked a little startled.
> "Nothing would please me better than to be taken out of this job. I want to go overseas. If they take me out of here, maybe I can get there."[17]

The congressman did contact the War Department, but the result was hardly what he expected. Instead of transferring Eisenhower, Assistant Secretary of War Benedict Crowell wrote a letter saying that Secretary Newton D. Baker had commended Eisenhower for his "diligence in looking after the welfare and well-being" of his soldiers. As for the hotel owner, there was nothing he could do but "come back, apologize, and renew his promise of good behavior." Although Eisenhower did not think he deserved a second chance, he relented, adding that the owner could expect provost guards dressed in civilian clothes to "be haunting your place," and that if they observed the sale of liquor to any man in uniform, "I'll put you out of business." That settled the matter.[18]

Some officers also posed problems for the harried young captain. One lieutenant, caught cheating at cards by his fellow officers, lied to Eisenhower when confronted with decks of cards he had marked. Infuriated, Eisenhower offered him the choice of resigning or facing a court-martial. The officer chose the former course.

Several days later the man's father, accompanied by his congressman, paid Eisenhower a visit. The congressman explained that the officer's father was one of his most important constituents and asked Eisenhower to withdraw the resignation and send the man to another camp. This Eisenhower refused to do, saying that he would not pass a problem on to another commander. Nor would he agree to deleting the words "for the good of the service" from the resignation. Although he heard no more from the congressman, he learned later that the man was reinstated.[19]

On 6 June the men at Camp Colt got their first look at a real tank when a French-built Renault light tank was delivered to the post.[20] Eisenhower later recalled that the event was marked by the "cheerful cynicism of [the] soldiers," who "had not expected to see one until we reached Europe." Apparently, he thought, someone in the war zone decided "there might be virtue in letting the Tank Corps recruits and trainees get a preliminary look at the machine that one day we would operate." This tank was later followed by two more, along with two British tank officers who served as advisers. The tanks, unfortunately, came without armament, but as always, "we improvised."[21]

June also brought the first call for shipment of Tank Corps volunteers overseas. Ironically, the request was for sixty-four telegraphers—not tankers. "This request gave some of us around headquarters a bit of sardonic amusement," Eisenhower wrote later. "We had been told that the Tank Corps in Europe had no need of the specialized training we had set up."[22]

Tank Corps volunteers continued to pour into Camp Colt during the late spring and early summer of 1918. By July, the 306th Heavy Tank Battalion and the 331st, 332d, and 333d Light Tank Battalions had been added to the post's rolls.[23] The sprawling encampment's population had grown to several thousand—all commanded by a twenty-seven-year-old captain. Welborn had been trying to get Eisenhower promoted to major since the middle of April. Although Eisenhower was only three years out of West Point and not eligible for promotion under current War Department rules, Welborn sought to justify the appointment based on Eisenhower's performance and need for rank commensurate with his responsibilities. No action was taken, however, until it was discovered that a bureaucratic ruling was at fault. It seems that Welborn's requests had failed to indicate that a vacancy existed for an officer in the grade of major. A request with the required wording was quickly submitted, and orders were cut on 22 July appointing Eisenhower as a temporary major with a date of rank of 18 July.[24]

Although all of the men at Camp Colt were volunteers exhibiting "exemplary zeal" and were "a few cuts above the average," there were a significant number who would be considered mentally retarded by today's standards. Rather than discharge such men, the War Department directed camp commanders to create "Development Battalions," the purpose of which was to prepare them for useful service. Eisenhower gave command of his Development Battalion to a former classmate, Capt. Norman Randolph, and transferred several hundred troops to the unit. Randolph, whose approach embraced the Salvation Army's philosophy "that a man may be down but he is never out," was so successful that eighteen graduates of his training program went on to earn commissions, and many others went on to become senior noncommissioned officers and technicians.[25]

Even with the beginning of overseas shipments of tank troops in August, Eisenhower had his hands full at Camp Colt. Three battalions, the 302d, 326th, and 327th, were transferred to Camp Tobyhanna near Scranton, Pennsylvania, in July and then overseas in September. While at Tobyhanna, the training of the three battalions was managed by Col. William H. Clopton, Jr., and a cadre of officers and NCOs who had received training in France and England.[26] Like Eisenhower at Camp Colt, Clopton was limited to providing training in basic soldier skills, with some theoretical instruction on the business of tank warfare thrown in.

The 303d, 328th, 329th, 330th, and 331st Battalions left Colt directly for France in August, followed by the 305th and 332d in September. They were replaced when the 334th through 339th Battalions were activated in August and September, and the 346th in October. Three of these, the 335th, 336th, and 337th, joined the exodus to France after just one month of training.[27]

Because Camp Tobyhanna was "manifestly unfit for winter training," Clopton and his staff closed down their operation there and moved to Camp Polk near Raleigh, North Carolina, in early October. They were joined by the officers and men of one of the heavy tank battalions at Camp Colt, who formed the nucleus of the training center, "which was to be augmented later by recruits received from a vigorous recruiting campaign then in progress."[28]

The Tank Corps reached its maximum strength with the activation of the 340th through 343d Tank Battalions at Camp Polk in early November. None of these units ever made it to France.[29]

The success of tank operations in the St. Mihiel and Meuse-Argonne campaigns had so impressed the Army staff that plans were made for the further activation of five more tank brigade headquarters elements (the 311th through 315th), four heavy tank battalions (the 309th through 312th), five light tank battalions (the 347th through 351st), and six heavy and five light training and replacement companies, seven repair and salvage companies, and a depot company. The war's sudden end killed this action, however.[30]

Providing experienced leaders for these units was impossible; there just were not enough to go around. First lieutenants commanded battalions and second lieutenants companies. Nor were there sufficient noncommissioned officers. An interesting (by today's standards) personnel action in which Eisenhower had a hand can be found in the files of Company C, 304th Heavy Tank Battalion.

Pvt. Paul R. Grant, a soldier in Company C, had been interviewed by 2d Lt. R. M. Marshall, the commander of Company B, 338th Light Tank Battalion, and had been accepted for duty as first sergeant—the senior enlisted position in the company. The action was further recommended by the 338th's commander, 1st Lt. C. R. Purcell, and Grant's company commander in the 304th, 1st Lt. Leonard Murphy. Eisenhower subsequently approved it.[31]

In September, when the camp's population reached its peak of more than 10,600 officers and men, Eisenhower faced a major crisis. The "Spanish flu," which entered the United States through the port of Boston and was first diagnosed at Fort Devens, Massachusetts, on 13 September, reached Colt on the fourteenth with 124 enlisted men shipped there from Devens for training as machine-gunners. Several of the men complained that they were ill, but the medical officer accompanying them, 1st Lt. T. J. Ferguson, diagnosed their condition as "a touch of the grippe." Ferguson left the train and went to a hospital in Lancaster, where he died several days later.

Eisenhower's own medical officer compounded the problem when he examined the men and decided that they were experiencing aftereffects from typhoid shots they had received before leaving Devens. The men were then assigned to a casual company to await training. Within a few days several of the Devens troops were hospitalized, and it was finally decided that they had contracted Spanish influenza.

Lt. Col. Thomas Scott, the camp's chief surgeon, reacted quickly to the emergency. He ordered the men isolated and began an experimental course of treatment involving the use of several "strong sprays."[32] The treatment was both curative and preventive, and Eisenhower, his family, and several members of the headquarters staff participated in the experiment. Recalling the episode later, Eisenhower wrote:

> Each morning he would use two sprays on the throat and nostrils of each of us. One of them was intensely pungent and strong. On application I felt as if the top of my head was going off. The other was, I think, a sort of soothing syrup to follow the first. The twice daily spray was anything but a tonic. Nevertheless, he insisted upon continuing this treatment on [all of us]—though possibly he spared the baby. We were fortunate—or he was smart. Not a single person in my headquarters command or my family contracted the flu.[33]

Eisenhower ordered daily medical examinations for every soldier in the camp, and any man showing symptoms of the illness was sent to one of the five infirmaries there. More severe cases were treated in special hospital tents. The cooperation of Gettysburg's citizens was also enlisted. Restaurants were ordered not to serve soldiers, who were also forbidden to attend church services in town. The local Roman Catholic church offered the use of its school, St. Francis Xavier Hall, for patients from Colt.

Between 15 September and 5 October a total of 321 cases of Spanish influenza were treated at the camp hospital, and 150 soldiers died. Another 106 soldiers were treated for pneumonia.[34] The performance of Colt's medical staff so impressed the War Department that thirty doctors were ordered away temporarily to show medical officers at other camps the procedures they had followed.[35]

On 14 October, Eisenhower's twenty-eighth birthday, he was promoted to temporary lieutenant colonel. With the promotion came orders to close the camp and ship the remaining troops to Camp Greene, North Carolina. This did not bother Eisenhower, as the move was not to be made until after the November overseas troop shipment, scheduled for the eighteenth, which Welborn had promised him he would command.[36]

Eisenhower's experience with the Tank Corps in the United States was particularly frustrating for him. He had excelled in every way, showing great ingenuity in developing a meaningful training program despite the lack of resources. He had also shown a great deal of skill in dealing with politicians and community and business leaders. He had been exposed to working with Allied officers, in the form of the two British advisers, and had been impressed by their knowledge and skill. Welborn was sufficiently impressed with his performance to recommend him for award of the Distinguished Service Medal, which he did not receive until 1922. But, more than anything, Eisenhower wanted combat duty, and nothing was going to keep him from it now that he had the opportunity.

Although Welborn had finally made good on his earlier promise to send Eisenhower overseas, he still tried to talk him out of going, offering him promotion to colonel as an incentive to stay. Eisenhower declined, saying, "I'm ready to take a reduction in rank to the average of my class—to major that is—if the lieutenant colonelcy which I have now stands in the way of my going overseas."[37] It proved to be an unnecessary offer—the Armistice settled the matter for them.

Notes

1. Eisenhower, *At Ease,* 137.
2. Ibid.
3. Ibid., 138.
4. Ibid., 139; *Order of Battle,* 1544–46.
5. Eisenhower, *At Ease,* 138–39.
6. Ibid., 139.
7. Ibid., 140.
8. Ibid.
9. *Order of Battle,* 1545–46.
10. George Martin, "Picking the Right Men for the Tanks," *Treat 'Em Rough* (Camp Colt, Gettysburg, Pa.), 29 May 1918, Col. Charles H. Phillips file, World War I Survey Collection, Archives, USAMHI, 1.
11. As quoted in *301st Battalion History,* 185–86.
12. Capt. A. B. Custis, Memorandum "By Order of Captain Eisenhower," 20 June 1918, Modern Military Records Division, Record Group 120, Entry 1299, Box 24, 302d Tank Battalion Document File, NA.
13. As quoted in Stephen E. Ambrose, *Eisenhower: Soldier, General of the Army, President-Elect, 1890–1952,* Vol. 1 (New York: Simon and Schuster, 1983), 63.
14. Capt. A. B. Custis, Memorandum on "Letter Writing . . . by Order of Captain Eisenhower," 12 July 1918, Modern Military Records Division, Record Group 120, Entry 1299, Box 24, 302d Tank Battalion Document File, NA.
15. Capt. Dwight D. Eisenhower, Memorandum to Commanders, 23 Apr. 1918, ibid.
16. Eisenhower, *At Ease,* 142.
17. Ibid., 144–45.
18. Ibid., 145.
19. Ibid., 143–44.
20. Merle Miller, *Ike the Soldier: As They Knew Him* (New York: G. P. Putnam's Sons, 1987), 169.
21. Eisenhower, *At Ease,* 146–47.
22. Ibid., 146.
23. *Order of Battle,* 1545–46.
24. Miller, *Ike the Soldier,* 171.
25. Eisenhower, *At Ease,* 141–42.
26. Brig. Gen. Samuel D. Rockenbach, "Report of Chief of the Tank Corps," 13 Oct. 1919, USAMHI.
27. *Order of Battle,* 1545–46.

28. Rockenbach, "Report of Chief."
29. *Order of Battle,* 1546–47.
30. Chart accompanying Memorandum for Operations Division, Office of the Chief of Staff, 30 Nov. 1918, Modern Military Records Division, Record Group 165, Entry 310, Box 447, File 7-61.13, NA.
31. Pvt. Paul S. Grant to commander, Camp Colt, 26 Sept. 1918, Modern Military Records Division, Record Group 120, Entry 1299, Box 29, Company C, 304th Battalion, Tank Corps, Document File, NA.
32. Miller, *Ike the Soldier,* 172.
33. Eisenhower, *At Ease,* 150.
34. Miller, *Ike the Soldier,* 173.
35. Eisenhower, *At Ease,* 150.
36. Miller, *Ike the Soldier,* 173.
37. Eisenhower, *At Ease,* 151.

CHAPTER V

Tank Production:
Made the American Way

While Rockenbach and Patton worked to establish a training base for the Tank Corps in France and England, other officers took the lead in setting up a production program that would provide them the tanks they needed. Unfortunately, their efforts would not be as successful.

As previously noted, Majs. James A. Drain and Herbert W. Alden, mechanical engineers detailed by the chief of ordnance in Washington to conduct a fact-finding mission on tanks in the fall of 1917, determined that the fastest way to equip the AEF Tank Corps would be to license American firms to build the Renault light tank. They also determined that none of the existing heavy tank designs was satisfactory and recommended that the British and Americans work out a new design and enter into an agreement to jointly produce it. This could best be accomplished, they said, if the United States provided the engines and other automotive parts, and England the armor plate, weapons, and ammunition. To speed up the process, they further recommended that an assembly plant be established in France, preferably at a coastal site near a major port and rail center.[1]

Providing labor for such a joint venture did not appear to be a problem to Drain and Alden. Because the work of assembling the components was considered unskilled, they were able to obtain a tentative agreement from the British to provide Chinese workers. Whatever foremen or other skilled laborers were required could come from England or the United States, with necessary training to be provided in British assembly plants.[2]

Pershing approved their recommendations in mid-November 1917, and Drain was apppointed to represent the United States on the Inter-Allied Tank Commission created that month.[3] His first task was to meet with his British and French counterparts and work out a mutually agreeable heavy tank design. Working closely with the British, Drain was able to come up with a working concept. The French, while supportive, showed no interest in participating in the project, choosing instead to rely on their existing tank designs.[4]

Gen. Tasker H. Bliss cabled the War Department at Pershing's request on 6 December, seeking approval for the production of 1,500 Mark VIII Liberty tanks by 1 October 1918. If approved, the project would require the United States to provide 1,500 Liberty (from which the tank's name was derived) twelve-cylinder aircraft engines, starters, clutches, radiators, fans and piping, silencers, electric lighting, dynamos and batteries, propeller shafts, transmissions, brake systems, and suspension components.

The British would provide the tank's armor plate, structural members, track shoes, guns and mounts, ammunition racks, and ammunition. Bliss further stated that the British promised to allocate the steel required to produce their components but expected "complete guarantees" that the United States would replace the high-grade steel used for the tank's armor.[5]

Drain, working closely with British tank experts, was able to complete work on the tank's design specifications later that month. A wooden model was built, complete in every detail, and presented, along with the plans, to Rockenbach and the AEF's chief ordnance officer in Paris on 26 December for approval.[6]

The design called for a vehicle thirty-three feet long; twelve feet wide; and nine feet, ten inches high. Weight was estimated at thirty-five tons. The tank's armor was to be of sufficient thickness to protect the crew and internal components from all small-arms bullets, including armor piercing. The Liberty engine was expected to provide a power-to-weight ratio of ten horsepower per ton and a top speed of six miles per hour. Armament would consist of seven machine guns and two six-pounder cannon mounted in retractable sponsons.[7]

After receiving approval for the project, Drain and his British counterpart on the tank commission, Col. Albert G. Stern, selected Neuvy-Pailloux near Châteauroux as the site for the assembly plant.[8]

The War Department sent word on 17 January 1918 that an initial shipment of Liberty engines for the Mark VIII would arrive in France in April, and possibly as early as March.[9] Armed with this information, the Inter-Allied Tank Commission hammered out a detailed agreement covering all the points specified in Bliss's 6 December 1917 cable.[10]

The first indication that there would be problems in the heavy tank program came a month later, when the War Department notified the AEF that the initial shipment of 200 Liberty engines would not arrive until May. An additional 300 engines were promised each month thereafter.[11]

Delivery of the engines was not to be. The Signal Corps, proponent for the fledgling Air Service, was siphoning away Liberty engines as fast as they came off the production line—which was not fast enough for any of the parties concerned. The Liberty, which the Allies thought to be one of the best aircraft engines designed up to that time, required numerous minor modifications before full-scale production could begin.[12] On 29 April, Gen. Peyton C. March, the Army chief of staff, notified Pershing that the best he could hope for was enough engines and other components to complete 365 Mark VIIIs for the AEF by the end of August, followed by an additional sixty tanks per month. To make matters worse, the French, who had initially said they were not interested in the heavy tank project, demanded that the clause providing for shipment of the first 600 tanks to the Americans be dropped and that an allocation agreement be worked out between the three governments. March wanted to know whether, if deliveries to the AEF were cut to the numbers cited, a production rate of an additional 135 tanks in August, 240 per month over the next three months, and forty in December would be enough to satisfy the British and French.[13]

That question became irrelevant when the War Department informed Pershing on 29 June that the number of engines available for tanks had again been cut. It now appeared that the best the AEF could hope for would be 100 engines by 1 October, an additional 100 that month and again in November, 125 in December, and 175 in January 1919.[14] This last cable led to Rockenbach's decision to beg for enough tanks from the British to equip the heavy tank battalions then training, or scheduled to train that summer, in England.

Efforts to obtain Liberty engines and other components from the United States during the remainder of the summer and early fall of 1918 proved fruitless. On 27 November Pershing finally cabled the War Department to report that the British and French no longer desired to participate in the program. Construction of the assembly plant at Neuvy-Pailloux, which had progressed at a snail's pace because of the delays, was suspended on the eve of its completion. "No tanks will be assembled in France," Pershing stated tersely. "Project terminated and assets will be liquidated."[15]

The attempt to produce an American-built copy of the Renault light tank for the AEF was slightly more successful but infinitely more frustrating for the officers involved in the project. Officers from the Ordnance Department met with representatives of the Renault Works in late 1917 to work out an agreement for production of the company's light tank in the United States. A bargain was struck, and the French promised to provide detailed plans as well as two production models of the vehicle for shipment to America. Little work was accomplished on the project, however, until Lieutenant Braine was ordered to report to Lieutenant Colonel Drain in Paris in January 1918. On the twenty-ninth, on Patton's recommendation, Rockenbach decided to send Braine back to the United States to act as the AEF Tank Corps's liaison with the light tank production effort there. Although the French had promised two copies of the Renault for shipment to America along with the tank's plans, all Braine was able to secure from the company were a turret, a 37mm cannon, and gun mounts.

Braine, plans in hand, was able to book passage for himself and the equipment on the USS *Apples,* an old German freighter that had been interned on the island of Haiti, and left from St. Nazaire on 22 February.[16]

Upon arrival in New York on 13 March, Braine encountered the first of a series of bureaucratic obstacles that would plague the light tank production effort throughout the remainder of 1918. After obtaining permission to load his equipment onto a tug for transfer to a dock, Braine spent most of the day searching for a berth that would accept his cargo. "After much running around and telephoning," he wrote later, "I got permission to [transfer the cargo to] the Ordnance Dock at Governor's Island." There were several times when it looked like his mission would end in failure as the precious shipment "almost landed in the bottom of the bay."[17]

Braine wired the chief of ordnance in Washington to announce his arrival and received instructions to report to Lieutenant Colonel Alden, who had been promoted and sent back from France to supervise tank production for the Ordnance Department. He left that same night and, after briefing Alden on the purpose of his mission and leaving blueprints for the 37mm gun with him, returned to New York the evening of the fourteenth. Alden further instructed him to ship the 37mm gun he had brought from France to Washington so it could be studied and incorporated into a turret design being prepared by engineers there. Braine was then to take the rest of his equipment and mechanical drawings and report to the Ordnance Department's Motor Equipment Section in Dayton, Ohio.[18]

Although the War Department had asked for civilian manufacturers to bid for the rights to construct the Renault light tank, several companies confused the issue by responding with design proposals of their own. The Ford Motor Company produced designs for a three-ton two-man tank and a seven-and-one-half-ton three-man model. The Endicott-Johnson Company, a shoe-manufacturing firm, financed a flame-throwing tank design, and the Pioneer Tractor Company designed and built a tank made from iron pipe and plumbing fittings.[19] While these firms continued to press on with their work, owing in large part to a lack of coordination in the Ordnance Department, the Maxwell Motor Company and several other firms were contracted to produce 4,440 light tanks based on the Renault design at a cost of $11,500 each.[20]

Confusion over the design of the turret for the Renault tank was one of the main reasons for the project's initial delay. Working independently of each other, Alden's office in Washington and officers in the Ordnance Department's Engineering Bureau in Dayton were preparing separate plans for the construction of the turret. The Ordnance Department finally cabled the Engineering Bureau in Dayton to cease further work on the turret and wait for the plans from Washington.[21]

All of this was unknown to Rockenbach and Patton in France. They were operating on the assumption—based on information provided by the Ordnance Department—that the first shipment of 100 light tanks would arrive in April, followed by 600 per month beginning in May. The only problem they were aware of concerned armament for the tanks. The War Department had asked Pershing to secure 2,400 Hotchkiss machine guns from the British and 1,600 37mm guns from the French with which to arm the tanks when they arrived in France.[22] Braine had been notified of this before leaving for the United States and was considerably dismayed when he found that not so much as a single "line [had been] drawn on paper before the first week in January."[23]

Following his initial meeting with Ordnance Department officers in the Engineering Bureau in Dayton, Braine was ordered to report to Col. Lucian B. Moody in the Ordnance Department in Washington. There a battle was fought over which section would have Braine's services. After considerable debate, he was finally attached to Colonel Moody's office, which was in charge of the Ordnance mobilization program for tanks and artillery, and instructed to devote whatever time he could spare to the other departments involved in design of the light tank.

Despite the optimistic report sent to France in February, Braine found that little actual work had been accomplished. No tools, spare parts, or other supporting equipment had been ordered. Nor had a contract for production of the 37mm gun been drawn up. When he discovered this, Braine insisted that the 37mm gun and mounts he had brought back from France be copied and put into production. His recommendation was over-ruled by Lt. Col. Earl McFarland, director of the Ordnance Department Engineering Division's machine-gun and small-arms section, who said they would use a modified version of the American 37mm field gun, for which they would design a special armor with a splashproof mask to protect the crew.[24]

Problems were also encountered with machine guns for the light tanks. Braine reported that the Ordnance Department was able to locate 7,206 Marlin-Rockwell aircraft machine guns in Fairfield, Ohio. Someone suggested that the guns could be modified for use on tanks simply by equipping them with pistol grips and, as with the 37mm guns, designing a special splashproof mask for the turret. Because the guns belonged to the Signal Corps, the Engineering Division sent a memorandum authorizing that the guns be purchased from that branch and the required modifications be made by the company.

Although Braine again was not consulted, he took special pains to see that the guns were shipped to the company's plant, in response to a cable from Washington. Three times he received word that the modifications had been made, only to find that no work had been accomplished. The company finally agreed to modify fifty guns by a specified date. Again the Marlin-Rockwell people failed to comply, this time offering the excuse that the Ordnance Department had failed to order the work in writing.

Braine then turned to the Ordnance Department's Supply and Procurement Division, where he was informed that no contract had been let for the project as an agreement on price could not be reached. In addition, he was told that the Requirements Division had asked that the guns be delivered over a period spanning both 1918 and 1919. Braine met personally with the head of the Requirements Division and persuaded him to authorize purchase of all the guns in 1918. The head of the Requirements Division then instructed Braine to write a letter to the Procurement Division explaining the reasons for the change so the matter could be settled.

Braine dictated a letter for the Requirements Division director's signature on 1 June. After initialing it, he left it with the division's assistant director and departed for Dayton. When Braine returned to Washington later that month, he found that the guns still had not been shipped. Upon investigation, he discovered that the assistant director had rewritten the letter and then left it on his desk without taking further action. All of this took some three and a half months, and, in the end, the Marlin machine guns proved to be unsatisfactory for use on tanks. [25]

Another example of how bureaucracy thwarted the light tank production effort can be found in the specifications for the speedometer. Braine had specifically requested that the speedometers be calibrated to read in kilometers per hour. Someone in the Ordnance Department had decided that the instrument should record speed in miles per hour and that the odometer should read in miles and tenths of miles.

Production of the speedometers was delayed when the Stewart-Warner Company, faced with conflicting requirements, cabled to ask how the instruments should be built. A compromise was finally reached, with the speedometers being made to record speed in miles per hour and distance in kilometers and meters. [26]

Braine complained bitterly about the lack of cooperation he received from the various sections of the Ordnance Department. Much of his time was wasted by "people who would send you from one person to another before you could get any information," a fact he found "very discouraging." Even more distressing, Braine was rarely informed of telephone calls made to him; and all of his mail, both official and personal, was opened before he received it. Furthermore, he was instructed not to write or cable his superiors in France except to discuss matters of a personal nature—"and then was repeatedly warned against" even this innocent gesture. His Ordnance Department supervisors "emphatically informed me that [the] people in France were fully advised as to the progress and situation at all times." [27]

But Patton and Rockenbach were not kept informed. They remained blissfully unaware of the bureaucratic nightmare Braine had encountered in the United States and did not awaken to the reality of the situation until a cable was received at AEF headquarters on 20 June informing them that only twelve light tanks could be delivered by 1 September. An additional 100 tanks were promised by 1 October, with deliveries to increase until they reached a peak of 500 tanks per month. [28] It was this cable that led to the decision to seek enough Renaults from the French to equip Patton's two trained light tank battalions in time for the St. Mihiel offensive.

After spending several months traveling back and forth between Dayton and Washington, "averaging twenty-two nights a month on a sleeper," Braine learned that Colonel Welborn had been appointed director of the Tank Corps in the United States. He announced his intent to visit with Welborn but was advised not to by Colonel Moody and Brig. Gen. John H. Rice, chief of the Engineering Division. Eventually, however, Braine had the opportunity to contact Welborn's office, "and when Colonel [William H.] Clopton [Welborn's deputy] arrived [he] was of great assistance to me in having someone to go to and tell my troubles."[29]

Exactly what troubles Braine discussed with Clopton are never specified. Certainly he addressed the bureaucratic problems he encountered, but it is also possible to speculate that he was concerned about the integrity of his Ordnance Department supervisors. While Braine is careful not to raise this issue directly in his narrative, the charge that they were withholding information about telephone calls placed to him, and that they were censoring his personal and official mail is a serious one. Even more serious is his claim that he was ordered not to communicate with his superiors in France—especially after they assured him they were providing Rockenbach and Patton accurate reports on the status of the light tank production effort. This was clearly not the case, as cable traffic between Washington and the AEF shows.

Another of Braine's major complaints was the lack of personnel involved with the project. In the late spring of 1918 he finally obtained permission to recruit twenty-five officers from the Tank Corps to assist with the work. Unfortunately, before Braine could go to Camp Colt in Pennsylvania to select these men, he was ordered back to Dayton. The task fell on other officers from the Ordnance Department, who "did not get the proper men for the job."[30]

In Dayton, Braine

> followed up [with] the Signal Corps with reference to the wireless apparatus, interphones, splash-proof face guards, steel helmets, and had experiments tried with triplex glass for eye slits, which proved successful. There were five or six interphone manufacturers with whom I took this matter up, and any number of equipment manufacturers [for] face guards, leather helmets, steel helmets, etc., but it was impossible to get anyone to make a decision on any of these matters.[31]

More than twenty independent contractors were involved in production of the Renault tanks, a fact which almost certainly contributed to delays.

The Maxwell Motor Company and the C. L. Best Company, both of Dayton, and the Van Dorn Iron Works of Cleveland were the major contractors. These three firms were responsible for tank assembly and manufacture of some integral parts. Other corporations were contracted to supply engines, transmissions, tools, magnetos, armor plate, radiators, and other parts, all of which "would necessitate a Philadelphia lawyer to keep track" of them.

Although frequent meetings were held, Braine said they usually hindered rather than helped the effort because "one manufacturer would see that the other man was not very far ahead, so he would see no occasion to hurry," and all would then contribute to further delay of the project.[32]

Conflicting lines of authority among various government agencies also came into play. Braine cited as an example a contractor with an order for a steel pin requiring a certain percentage of carbon. The contractor told the Engineering Division that he could begin work if he could use steel that was readily available with a little less than the required carbon. He was advised he could do this. However, to do so, he first had to submit a new drawing to a government inspector working for a different division. When the material was not passed by the inspector, the district chief refused to approve the account, which meant the contractor was not paid — even though he had received prior approval to use the other material from the Ordnance Department's Engineering Division.[33]

Braine encountered similar problems in his effort to obtain tools for the tanks. After he drew up a list of the required tools, Braine was advised that he would need to have a draftsman prepare drawings of them so they could be purchased from private contractors. Once this work was accomplished, Braine took the drawings to Washington, where he was told that they had not been done on the proper form. After the drawings were redrawn on the required form, officers in the Procurement Division told Braine he would have to substitute standard Ordnance tools wherever possible. Braine went through the Ordnance tool catalogs and, after identifying all those that could be substituted, placed his order. Several weeks later he was informed that most of the Ordnance tools were out of date and that he would have to resubmit his original drawings for competitive bidding by private contractors. This meant an additional month-and-a-half delay.[34]

In the early summer of 1918, thanks to the efforts of a friend in the New York City Recruiting Office, Braine was able to indirectly contact Benedict Crowell, the assistant secretary of war. Crowell, according to Braine, took a personal interest in tank production, and spent five days investigating Braine's charges. At about this same time, Lieutenant Colonel Drain returned from Paris and, after reviewing the situation, played a key role in securing the appointment of Louis J. Horowitz as the civilian head of tank production, a move that Braine had earlier recommended. Horowitz's appointment, according to Braine, contributed significantly to breaking up the bureaucratic logjam.[35]

While Braine battled with the bureaucrats in Washington and Dayton in an effort to speed up production of the Renault tank, the Ford Motor Company pressed on with its own light tank design. A number of officers in the Ordnance Department were impressed with the Ford model and especially with Ford's promise to quickly mass-produce the vehicle.

Braine was invited to Detroit in the late spring of 1918 to see a prototype model of the three-ton two-man tank. He was not impressed. The Ford people were unable to get the engine, a complicated affair that consisted of two standard Ford motors hooked up in tandem, to start. In addition, the tank lacked a tailpiece, which Braine had learned from his experience in France was needed for trench-crossing operations. He later learned that the Ordnance Department notified his superiors in France that he approved of the design, "which was not so."[36]

On 9 August, the Ordnance Department cabled Pershing a glowing report extolling the virtues of the Ford tank. The department highlighted the fact that the machine's thirty-six-horsepower engine developed eleven horsepower per ton, the highest power-to-weight ratio yet for a tank, and that it could attain a top speed of ten miles per hour. Ordnance officers thought the tank was at least as good as the Renault, and, best of all, it could be produced in one-fifth the man-hours required to build its competitor. The department also informed Pershing that a sample Ford tank had been shipped to France on 7 August and would be in the hands of the AEF's chief ordnance officer by the end of the month.[37]

On 8 November, after Tank Corps officers had had a chance to test the Ford vehicle, Pershing cabled a recommendation to the War Department that the Ford be built for use as a light artillery tractor, as it did not have "sufficient value as a Tank to justify its production for that purpose except as an emergency substitute." Pershing then demanded that production of the Renault "be pushed and that no interference with its production be permitted," as his tank experts considered it to meet the minimum size and power requirements for a light tank.[38]

Pershing's protests fell on deaf ears, however, as the War Department had contracted with Ford several months earlier to produce 15,015 three-ton light tanks at $4,000 per copy. By war's end, despite Ford's highly touted assembly-line prowess, only fifteen had been completed. The remainder of the order was subsequently cancelled.[39]

By early fall the Renault production line finally began to move, and the first light tanks began rolling off in October. Braine, anxious to get back to France, encountered resistance to his departure from the Ordnance Department, so he turned to Colonel Welborn, who issued orders for his return that same month.

On 20 November, almost nine months to the day after Braine left France on the USS *Apples,* and nine days after the Armistice was signed, two American-built Renaults arrived at Bourg. Eight more light tanks followed in early December, bringing to ten the total delivered to the AEF.[40]

The failure of the Ordnance Department and American industry to provide tanks for the AEF Tank Corps was not an isolated instance. Similar failures were encountered in the production of trucks, airplanes, artillery, and small arms for the AEF. In fact, except for four fourteen-inch naval guns, no American-made cannon or shells were fired by First Army artillerymen. The Air Service did a little better, managing to equip twelve of its forty-five squadrons with American-built aircraft before the Armistice was signed. While most of the doughboys carried Springfields or Brownings into battle, the AEF still had to buy 9,592 Hotchkiss machine guns and 40,000 Chauchat automatic rifles to meet their needs at the front.[41]

More than ten years later, in a letter to the chief of cavalry, the assistant director of the Army Industrial College wrote, after reading Braine's report, that it was

a fine example of [our] lack of industrial planning and a horrible commentary on our preparedness when we entered the last war. The records of our branches are full of this sort of data—unfortunately not as forcibly stated in many cases. It is on data of this sort that we are getting the details of things we must plan to avoid in our next war.[42]

Looking back, it is not surprising that events progressed as they did. While considerable thought had been given to the problem of mobilizing manpower in the years prior to World War I, virtually nothing had been done to mobilize industry for a major war effort. The archaic bureaucracy that had so badly managed logistical support in the Spanish-American War was still in place when America declared war on Germany in 1917. It required more than half a year to sort out organizational problems alone.

The Ordnance Department, which managed production of more than 100,000 line items for the Army, had only ninety-seven officers assigned when America entered the war. Personnel planners projected the Army would reach a wartime peak strength of 5,000,000 men and that 11,000 trained officers would be needed to oversee the Ordnance Department's efforts to equip them. Civilian industry was able to provide the engineers needed by the department, but it took time to train them in the special problems involved in the production of weapons and munitions.[43]

Nevertheless, we are still faced with the question of why Braine so forcibly stated his case. Certainly he understood the time that would be required to convert the Renault's specifications from the metric system to the English measuring system employed in American factories. Nor was the lack of planning, poor coordination, and bureaucratic bungling he encountered as he shuttled back and forth between Dayton and Washington likely to have been the chief cause of his anger. It is possible to speculate, however, that Braine was truly incensed over the manner in which his Ordnance Department superiors handled the decision to produce the Ford light tank. Were they motivated solely by interest in Ford's assembly line capacity and a desire to save the taxpayers' money? If so, why did they forbid Braine to communicate directly with his superiors? Why did they intercept his incoming calls and mail, and send false reports to AEF headquarters indicating he thought highly of the Ford tank design when "it just was not so," as Braine so emphatically stated?

Braine and his AEF Tank Corps counterparts clearly believed the Renault light tank was the superior vehicle. The thought that someone in the Ordnance Department might have been making a personal profit by pushing the Ford design at the doughboys' expense may have infuriated Braine. The thought might well have led him to write his scathing indictment of "the system," and to challenge, indirectly, the integrity of those who were responsible for its operation.

Notes

1. Drain and Alden, "Report on Investigations," 6.
2. Ibid., 7.
3. Rockenbach, "Tank Corps Operations," 2.
4. Ibid., 2–3.
5. Cable, Gen. Tasker Bliss to Army chief of staff, 6 Dec. 1917, included as enclosure to Collins, "Development of the Tank Corps."
6. Collins, "Development of the Tank Corps."
7. Rockenbach, "Tank Corps Operations," 4–5.
8. Ibid., 4.
9. Cable no. 652-R, 17 Jan. 1918, included as enclosure to Collins, "Development of the Tank Corps."
10. "Agreement Between the British and U.S. Governments for the Production of Tanks," 22 Jan. 1918, ibid.
11. Cable no. 816-R, 22 Feb. 1918, ibid.
12. Maurer Maurer, ed., *The United States Air Service in World War I: The Final Report and a Tactical History,* Vol. I (Washington, D.C.: Office of Air Force History, 1978), 82.
13. Cable no. 1201-R, 29 Apr. 1918, included as enclosure to Collins, "Development of the Tank Corps."
14. Cable no. 1636-R, 29 June 1918, ibid.
15. Cable no. 1929-S, 27 Nov. 1918, ibid.
16. Braine, "Personal Experience Report."
17. Ibid.
18. Lt. Elgin Braine to Lt. Col. George S. Patton, Jr., 28 Mar. 1918, Patton Chronological Files, Box 9, 25–30 Mar. 1918, Patton Collection.
19. Mildred H. Gillie, *Forging the Thunderbolt* (Harrisburg, Pa.: Military Service Publishing Co., 1947), 9.
20. Cable no. 1233-R, 3 May 1918, included as enclosure to Collins, "Development of the Tank Corps"; Benedict Crowell and Robert F. Wilson, *The Armies of Industry: Our Nation's Manufacture of Munitions for a World in Arms, 1917–1918,* Vol. I (New Haven, Conn.: Yale University Press, 1921), 196.
21. Braine, "Personal Experience Report."
22. Cable no. 722-R, 1 Feb. 1918, included as enclosure to Collins, "Development of the Tank Corps."
23. Braine, "Personal Experience Report."
24. Ibid.
25. Ibid.

26. Ibid.
27. Ibid.
28. Cable no. 1568-R, 20 June 1918, included as enclosure to Collins, "Development of the Tank Corps."
29. Braine, "Personal Experience Report."
30. Ibid.
31. Ibid.
32. Ibid.; Crowell and Wilson, *The Armies of Industry,* 196.
33. Braine, "Personal Experience Report."
34. Ibid.
35. Ibid.; Crowell and Wilson, *The Armies of Industry,* 199.
36. Braine, "Personal Experience Report."
37. Cable no. 1828-R, 9 Aug. 1918, included as enclosure to Collins, "Development of the Tank Corps."
38. Cable no. 1879-S, 8 Nov. 1918, ibid.
39. Crowell and Wilson, *Armies of Industry,* 196–197.
40. Table included as enclosure to Collins, "Development of the Tank Corps."
41. American Battle Monuments Commission, *American Armies and Battlefields in Europe: A History, Guide and Reference Book* (Washington, D.C.: U.S. Government Printing Office, 1938), 504.
42. Colonel Irving J. Carr to Major General H. B. Crosby, 23 April 1929, Patton Military Papers, Box 47, Personal Experience Reports of Tank Operations – 1918, Patton Collection.
43. Crowell and Wilson, *Armies of Industry,* 20–26.

Brig. Gen. Samuel D. Rockenbach, commander of the American Expeditionary Forces Tank Corps.

Lt. Col. George S. Patton, Jr., commander of the 1st/304th Tank Brigade.

Capt. Dwight D. Eisenhower poses in front of a Renault Char FT light tank at Camp Meade, Maryland. This picture was taken in 1920, after postwar rank reductions for regular army officers went into effect.

Maj. Sereno Brett, commander of the 1st/304th Tank Brigade's 326th/344th Tank Battalion. Brett was the only Tank Corps officer to continue working with tanks throughout the 1920s and 1930s.

Capt. Ranulf Compton, commander of the 1st/304th Tank Brigade's 327th/345th Light Tank Battalion.

Much of the gunnery training at the 311th Tank Center at Bourg was conducted from Renault mock-ups like these. Note the rocker on the bottom of the trainer. This allowed the gunner to simulate vehicle movement while firing.

Tank crews prepare to mount their Renaults for training at the 311th Tank Center at Bourg.

A platoon of Renault light tanks moves into line formation during a training exercise at the 311th Tank Center near Langres on 15 July 1918.

A tanker fires an 8mm Hotchkiss machine gun from a ground mount during gunnery training at the 311th Tank Center at Bourg.

Tankers from the 301st Heavy Tank Battalion watch a demonstration of the British Mark IV's capabilities at Bovington Camp near Wool, England.

"The race track" at the 301st Tank Center in Bovington, England. American tank crews took initial driver training with the huge British Mark IV and V heavy tanks at this facility.

Crewmen undergo mechanical training on a Renault Char FT light tank at the 311th Tank Center at Bourg in the summer of 1918.

The Mark VIII Liberty tank was to be an Anglo-American joint venture, but the project was terminated shortly before the Armistice. One hundred were built by the United States in 1919.

A 327th/345th Battalion Renault advances with 42d Division doughboys west of Essey during the St. Mihiel offensive.

A pair of French Schneider tanks near Neuvilly on 3 October 1918. The cumbersome Schneider was the French army's first combat tank.

French St. Chaumond tanks advance near Blerecourt on 8 October 1918.

PART II

"[I] was made to understand by the Colonel
that a Tank Officer was meant to die."
                                    —2d Lt. Julian K. Morrison
                                    Company A, 344th Tank Battalion
                                                        Fall 1918

CHAPTER VI

St. Mihiel: First Blood

The decision to launch an independent American army in an offensive to reduce the German salient in the vicinity of St. Mihiel in the summer of 1918 marked the end of more than a year of frustrating dealings with the French and British and the realization of a dream for Pershing. The AEF's commander in chief had fought an embittered battle with Allied leaders who wanted to parcel out American units in an effort to bolster the wearied ranks of British and French formations.

As early as June 1917 Pershing had seen the strategic value of an assault on the St. Mihiel salient. He figured that if the German line could be breached in this area, the success could be exploited by a deep thrust to capture Metz, then continue up the Metz-Thionville railroad to the Thionville-Longuyon-Sedan railroad. These railroads were vital to the German lines of communication, and their loss would threaten the entire German position in France.[1] Pershing and French Marshal Henri Pétain reached a tentative agreement that same month that the railroads should be an American objective. A strategic study completed in September 1917 by Col. Fox Conner and Lt. Cols. LeRoy Eltinge and Hugh Drum of the GHQ staff endorsed this view and recommended that it be the major AEF effort in 1918.[2]

Pershing's plans for raising an independent American field army were threatened by the British and French during the winter of 1917-18, however. They lived in fear of the possibility that the Germans might any day shift large numbers of troops from the Eastern Front to France. Faced with dwindling manpower, they saw the amalgamation of American infantry companies or battalions into their own units as the only means to counter this threat.

Because an army requires an enormous supply organization, British civilian and military leaders pleaded with Pershing to give up his plans for an independent American army and assent to amalgamation, thus freeing more space on precious shipping for combat troops. Pershing turned a deaf ear to their pleas, so both the British and French appealed to President Woodrow Wilson for a decision. Wilson and Secretary of War Newton D. Baker interpreted the question as a military decision and advised Pershing in December that he had full authority in the matter, although both thought that maintaining the integrity of American forces in France was "secondary to the meeting of any critical situation by the most helpful use possible of the troops at your command."[3]

Armed with this information, Pershing continued to resist the Allies. Pressure for amalgamation mounted in January 1918, however, when Secretary of State Robert Lansing provided the gist of the cable to the French ambassador, who in turn notified the British that both of Pershing's civilian supervisors were sympathetic to their cause. In the end, Pershing was forced to give in to their demands—but on his terms. The agreement reached by Pershing and British Prime Minister David Lloyd George called for the United States to provide six combat divisions for use in the British sector. These units would be transported to the Continent on British ships in excess of those previously committed to the Americans. This compromise satisfied both parties. The British got the troops they wanted (in all, ten American divisions trained with the British, of which four made it to the front, and two—the 27th and 30th in the American II Corps—remained in combat with the British for the duration), and Pershing was able to continue building up his own army.[4]

But the amalgamation issue had not been laid completely to rest. The French wanted their fair share. To satisfy them, Pershing allowed a number of American divisions to go into the line in the French sector (including the 1st, 4th, 26th, 28th, 42d, 77th, and 2d, which included the 4th Brigade made up of two Marine Corps regiments) and, later in the spring, transferred four black infantry regiments to the French to become organic parts of French divisions.[5]

On 10 July, Marshal Ferdinand Foch delighted Pershing with the announcement that, with a million Americans on French soil, it was time for the American army to "become an accomplished fact." The two agreed that the First Army should assemble in the vicinity of Château-Thierry within three weeks. This sector was chosen because Maj. Gen. Hunter Liggett's I Corps was already in action there, as were several other American divisions. All that would be required to make the First Army a reality would be the creation of an army headquarters and staff and the addition of Maj. Gen. Robert L. Bullard's III Corps to the Army's order of battle. Pershing accomplished the former with a general order issued on 24 July, and the latter by assigning Bullard a corps sector on 4 August.[6]

Initial plans called for the First Army to be committed in an assault across the Vesle River. But Foch lost interest in pressing the offensive in this sector, and Pershing was permitted to revive his plan for the reduction of the St. Mihiel salient. He would have to give up the idea of driving on to Metz and the rail lines beyond, however, because Field Marshal Sir Douglas Haig had come up with a plan to exploit the successes of the British Expeditionary Force in the north. Haig's plan called for a compressing envelopment of the German army, with the BEF driving east through Belgium and northern France and the Americans and French attacking north through the Meuse River–Argonne Forest region. In late August Foch accepted Haig's plan, which called for the American First Army and French Fourth Army to strike northwest toward Sedan and Mézières between the Meuse River and Rheims on 26 September; followed by a British Fifth Army drive on Cambrai on the twenty-seventh; an attack by the British Second Army, the Belgians, and several French divisions in Flanders on the twenty-eighth; and the British Fourth Army's push toward the Hindenburg Line near St. Quentin on the twenty-ninth. Pershing was authorized to commit the First Army at St. Mihiel, but only if he could redeploy his force to the Meuse-Argonne sector and be ready to go on the offensive there in late September. He agreed to this condition.[7]

It was a monumental task, requiring as it did planning for two major offensive operations plus the logistical problems of redeploying his army first to the St. Mihiel sector, then to the Meuse-Argonne. But Pershing accepted the challenge with relish. "I never saw General Pershing looking or feeling better. He is sleeping well. He is tremendously active," wrote Charles G. Dawes in his diary. "He will soon strike with his field army."[8]

Pershing took over the Vesle sector from Gen. Joseph Degoutte on 10 August and immediately issued orders to relocate First Army headquarters to Neufchâteau and to begin planning for the St. Mihiel offensive.

On 18 August Brigadier General Rockenbach was ordered to meet with Assistant Secretary of War Benedict Crowell in Paris, then to proceed to

First Army headquarters to begin planning for the Tank Corps's role in the upcoming offensive.

In Paris, Rockenbach and Crowell met with representatives of the British and French governments in an attempt to speed up delivery of the tanks the Americans had been promised in June. They agreed that Rockenbach should dispatch six officers from Bourg to supervise rail-loading of the light tanks at the French factories and accompany them to Bourg.[9]

That mission accomplished, Rockenbach reported to Colonel Drum, then serving as First Army chief of staff, on Tuesday, 20 August, for temporary duty with the headquarters at Neufchâteau. Rockenbach was briefed on the plan for reducing the St. Mihiel salient and told to develop his own plan for the employment of his two battalions with 144 light tanks, plus a French light tank regiment with three battalions bringing 225 more Renaults, and three British heavy tank battalions with 150 Mark V tanks that had been allocated to First Army. Drum told Rockenbach that the high command thought casualties would be excessive if the heavy tanks were not given the mission of breaking through the great maze of wire obstacles believed to be in the sector.[10]

The remainder of that day proved to be a busy one for Rockenbach. In addition to establishing a headquarters for the Tank Corps at Ligny-en-Barrois, he had to organize two subordinate brigade headquarters elements and to appoint commanders and order them to report for further instructions. The British and French brigade commanders and their staffs were also asked to report at once.

Rockenbach decided that Patton would be given command of the 1st Tank Brigade, which would consist of the two light tank battalions he had trained at Bourg. Rockenbach's chief of staff, Lt. Col. Daniel D. Pullen, was directed to assume command of the 3d Tank Brigade staff, which would serve as the liaison between First Army and the French tank regiment that was being attached for the operation. Just five days before, Pullen had been released to the 2d Engineer Regiment because he feared "he would never get into the fight if he waited for Tanks."[11]

When Patton arrived at Neufchâteau, Rockenbach informed him of his new responsibilities, briefed him on the operation, and sent him to reconnoiter the V Corps front with his reconnaissance officer, 1st Lt. Maurice H. Knowles. Pullen and his reconnaissance officer were told to check out the IV Corps sector.

Rockenbach had personally reconnoitered the area in January and February and determined that "only a very limited use of tanks was possible." He noted at the time that light tanks would have to be preceded by heavy tanks or dismounted engineers, as "the trenches were too wide and crumbling" for light tanks to make the initial crossings. He also observed that the water table in the area was very high and that it would require very little rain to create bogs. The Rupt de Mad, a tributary of the Moselle River running through the sector, "was an impassable obstacle at all times unless the bridges were intact."[12]

Patton's report following his personal reconnaissance the night of 21–22 August was far less pessimistic and reinforced his belief in the "absolute necessity for a tank officer to personally see the ground." While his initial observations conducted from forward positions tended to confirm earlier reports that the ground was "an impassable swamp further blocked by dreadful barbed wire," his participation in a French raid into no-man's-land convinced him that neither the ground nor the wire was a serious enough obstacle to rule out an attack by his light tanks.[13]

Patton discussed his findings with Rockenbach on Saturday the twenty-fourth. He described the terrain as a flat, marshy plain drained by numerous small streams and ditches. Foliage along the waterways offered cover and concealment for "prone lines of skirmishers," and Patton thought both infantry and tanks could cross the streams in the sector with ease. A recent dry spell had allowed the ground to harden, but Patton cautioned that two days of rain was all that was needed to make the "sector quite difficult to tanks." He added that if the attack were postponed until late September, the anticipated rainy season would inhibit trafficability even more and force tanks to operate solely on roads. The much-talked-about wire obstacles would pose little difficulty. Patton said some of the wire was strung from iron posts that could easily "be flattened by tanks," while the remainder was attached to wooden posts that had rotted after years of neglect. "It was possible to push these posts over by hand," he wrote.[14]

Patton recommended that the Americans follow the British example at Cambrai and forgo a prolonged artillery preparation. Such a barrage would only alert the enemy that an attack was forthcoming and would tear up the ground, restricting the movement of tanks. In line with the tactics he had proposed to Pullen in July, Patton suggested that artillery smoke be employed to screen the tanks from German guns.[15]

The British sent word the next day that they would be unable to provide heavy tank support to First Army. This news particularly distressed Rockenbach, as it meant the loss of 300 six-pounder guns (the equivalent of seventy-five batteries) and 600 machine guns—a significant reduction in firepower for covering the assault.

Rockenbach, after meeting with both Patton and Pullen to discuss their terrain reports, and armed with the information that the heavy tanks would not be forthcoming, submitted a plan to First Army based on a number of "ifs." He included a lengthy explanation of the limitations of light tanks. He also submitted a request for a reduction in the planned artillery preparation, as he agreed with Patton that the proposed thirty-one-hour shelling would only make the terrain more difficult for his light tanks to negotiate.

As the days passed and preparation for the offensive continued, Rockenbach reminded his superiors at every opportunity that "the Tanks available could not do what the original plan contemplated" and that nothing in the First Army's final attack plan should be based on the performance of critical missions by light tanks.[16]

By 27 August it was apparent that all 144 light tanks allocated to the two American battalions at Bourg would arrive there by 1 September and, if everything went right, could be ready for combat by the fifth. Lt. Col. Emile Wahl, commander of the French 1st Assault Artillery Brigade, was also instructed to report to Rockenbach with his staff on 27 August. Wahl, it was reported, would come with not three but four light tank battalions equipped with a total of 300 Renaults.[17]

This latest information allowed Rockenbach to finalize his order of battle and allocate forces to the First Army. The breakdown called for two of the French battalions to support I Corps on the right (east), the remaining two French battalions to support IV Corps in the center, and Patton's brigade to support V Corps on the left (west). A request was also submitted to GHQ on 30 August for the necessary orders to ensure that trains would be available by H-minus-144 hours so tanks could reach the IV Corps detraining point at Ansauville by H-minus-78 hours, and the V Corps detraining point at Sommediene by H-minus-54 hours.[18]

Patton, meanwhile, returned on 24 August to Bourg, where he found the men at the Tank Center working diligently under the watchful eye of Major Viner in an effort to get the new tanks, which had begun arriving that day, ready for combat. Viner, who had been commanding the 326th Tank Battalion, was told to take over the operation of the center and school. Patton then shifted Major Brett from command of the 327th Tank Battalion to the 326th, and gave Capt. Ranulf Compton command of the 327th.

That night Patton returned to the V Corps area accompanied by 2d Lt. George B. Heilner, the 326th's reconnaissance officer, and 1st Lt. Harry

G. Borland, the battalion's gas officer. He spent Sunday the twenty-fifth drafting his attack plan, while Knowles, Heilner, and Borland set up a brigade command post and reconnoitered the area in search of detraining and assembly areas for the battalions, sites for subordinate unit command posts, and routes to the front line. They also laid telephone wires from the corps headquarters to the sites selected for battalion command posts.[19]

On Wednesday the twenty-eighth, Patton completed a revised plan that showed "in detail the operation of nearly every Tank." Later that day he issued his field order, which provided a detailed exposition of the German troop dispositions, the V Corps battle plan, and his own personal instructions to the tank crews. The brigade adjutant, Capt. Edmund N. Hebert, and the tactical officer, 1st Lt. Harry E. Gibbs, arrived at 10:30 P.M., "But truck and clearks were lost."[20]

The next morning Brett and his three company commanders, Capts. Harry H. Semmes (Company A), Newell P. Weed (Company B), and Math L. English (Company C), met with Patton to discuss their duties in detail.[21]

On 30 August Patton met with Maj. Gens. Clarence R. Edwards, commander of the 26th Infantry Division, and George Bell, Jr., the 33d Infantry Division commander, to discuss the employment of infantry with tanks. Both officers were "interested" and "eager to help." To further facilitate harmonious tank-infantry operations, Patton arranged a demonstration by one of the battalions at Bourg for officers of the two divisions.[22] Ninety attended, and Patton wrote his wife that night, telling her that the performance had been "better than I could have hoped. One tank fell off a cliff and rolled over then went on again without any difficulty."[23]

A supply dump was established on 1 September at Ecrouves for the tanks that would operate on the southern face of the salient. The dump, although insignificant as far as First Army was concerned, contained 20,000 gallons of gasoline, 2,000 gallons of light automotive oil, 600 gallons of heavy automotive oil, and 600 pounds of grease—all of which were vital to the tanks' operation.[24]

Patton's plans were altered slightly on 2 September, when he was notified by V Corps of minor changes in the attack zones. He quickly made the appropriate adjustments but was forced to discard his entire plan when word was received the next day that the attack had been postponed from 7 September to 12 September, and that the 1st Tank Brigade was being withdrawn from support of V Corps and attached to IV Corps in the center sector.[25]

The decision had been made to postpone the attack because of problems the newly organized First Army encountered in assembling units, coordinating artillery support, and constructing support facilities for the offensive. All of Patton's hard work in preparation for the attack had been for naught, and he greeted the decision with "some profanity and much regret."[26]

Later on the third, Rockenbach received an update on the composition of the French tank force allocated to First Army. Instead of the four light tank battalions he had been told were coming, the French were sending the 505th Assault Artillery Regiment with three battalions of light tanks; the XI Groupement, consisting of two St. Chaumond–equipped groupes (company-sized elements); and the IV Groupement, made up of two groupes of Schneider tanks. Wahl would command the 505th Regiment and XI Groupement, maintaining liaison with the American I Corps through Pullen's 3d Tank Brigade headquarters, while the IV Groupement, commanded by Maj. C. M. M. Chanoine, would operate with Patton's 1st Tank Brigade.[27]

It is interesting to note that the total number of tanks committed in support of the St. Mihiel offensive amounted to 419 — a considerably larger number than the 267 usually cited by most historians. The breakdown included 168 in Patton's 1st Tank Brigade (144 French-built Renaults in the 326th and 327th Battalions, and 24 Schneiders in Chanoine's IV Groupement), and 251 in Pullen's 3d Tank Brigade (216 Renaults in the 505th Regiment, and a mix of 35 St. Chaumonds and Schneiders in XI Groupement).[28]

The St. Mihiel salient formed a large, triangular bulge in the Allied line, with the left flank anchored in the east on the Moselle River at Pont-à-Mousson. The German trenches then swung south and west along a line behind Seicheprey. The right flank was anchored in the west at Haudi-omont. At the apex of the salient was St. Mihiel. Manning this pocket were eight war-weary German divisions and a separate brigade. Arrayed against them was a force consisting of four French and eight and a half American divisions.

Pershing's revised plan called for I Corps on the far right (east) and IV Corps in the center to advance northward through the salient and link up with V Corps, which would attack eastward across the salient from positions on the western face. The French II Colonial Corps with three divisions had a follow-on mission in the vicinity of St. Mihiel at the apex of the triangle. I Corps was made up of the 2d, 5th, 82d, and 90th Divisions; IV Corps the 1st, 42d, and 89th; and V Corps consisted of a brigade of the 4th, the 26th, and the French 15th.[29]

Engineers constructed dams near Bouconville in the 89th Division's sector to prevent the Rupt de Mad from rising in the event of rain, and stockpiled materials in case it became necessary to repair the bridges needed to traverse the Rupt de Mad.[30]

Patton met with Pullen at Ligny and discussed the new situation as they rode together to IV Corps headquarters at Ecrouves. Pullen had already drawn up a tentative plan for the tanks, but Patton objected to the large frontage Pullen had allocated to the battalions.[31]

With the new attack plan in hand, Patton began a detailed reconnaissance of his new sector, which included the area from Bouconville to Bois de la Hazelle, on Wednesday, 4 September. His study of the area led him to recommend employment of his tanks in the sector between Marvoisin and the Beney-Pannes-Essey road. He was also able to convince the IV Corps chief of staff and G-3 to reduce the frontage assigned to his battalions.[32]

The 1st Tank Brigade was given the mission of supporting the 1st Infantry Division on the left (west) and the 42d Infantry Division in the center of the IV Corps sector. To accomplish this task, Patton allocated Major Chanoine's two groupes of Schneiders and Captain Compton's 327th Tank Battalion (minus sixteen tanks that would be held in brigade reserve) to the 42d Division. Major Brett's 326th Battalion would support the 1st Division.[33]

Patton gave careful consideration to his assessment of the strengths and weaknesses of his subordinate commanders, the limitations of their equipment, and the level of training of the units they were to support when he formulated his plan of attack. Brett, in whom Patton had complete confidence, was instructed to attack across the Rupt de Mad with his battalion, supported by the brigade reserve, and lead the doughboys of the 1st Division to their objectives. Patton foresaw no problems in this sector, as the 1st Division had operated with tanks at Cantigny and was familiar with their tactics. Chanoine's heavier Schneider tanks were expected to have a rough time crossing streams and other obstacles in the center sector, so Patton directed them to follow the infantry. Specifically, the 14th Groupe would operate in the 165th Infantry Regiment's sector, and the 17th Groupe with the 166th Infantry Regiment. To their right, Compton was instructed to follow the advance of the 167th Infantry Regiment initially, then to pass through and lead the way into Essey and Pannes.[34] The facts

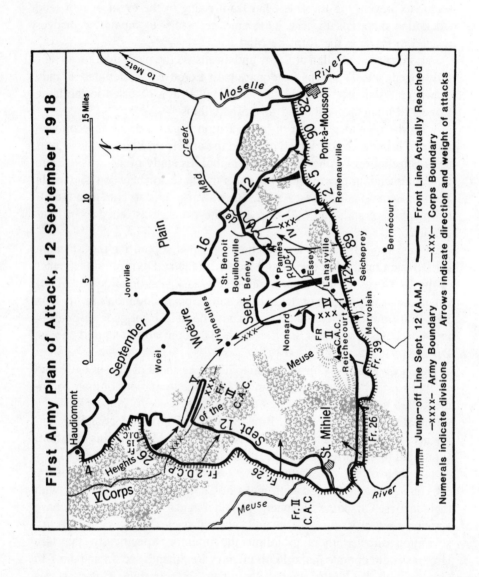

First Army Plan of Attack, 12 September 1918

that the 42d Division had no previous experience with tanks, that Patton was still uncertain of Compton's abilities, and that the two French tank groupes were to be deployed in the 42d's sector may have caused Patton to devote greater than normal attention to his preattack coordination with the 42d.[35]

While Patton was working out the details of his attack plan, Brett and Compton were busily engaged with the task of preparing their men and new tanks for battle. Thirty-six tanks were delivered to Compton's battalion on 27 August from the Berlier factory at Cercottes. An additional thirty tanks were delivered the next day, and the final six arrived on 3 September. Besides getting the tanks themselves ready to fight, Compton had his crews undergo a refresher course in the firing of the French 37mm Poteaux cannon and the 8mm Hotchkiss Model 1915 machine guns employed on their tanks. They also participated in company-level training conducted by commanders until 6 September.[36] Similar activities were conducted by the 326th Battalion during this period.

Prior to moving out, Patton directed that each tank be marked on the turret to facilitate battlefield command and control. The markings he chose were the playing card suits (hearts, spades, clubs, and diamonds). These were stenciled in black on varying shaped white backgrounds. The suit served to identify the platoon, while the background denoted the company to which the vehicle was assigned. A numeral (one through five) was stenciled next to the unit marking to identify each tank within a platoon. For example, the platoon leader's tank in Company C, 326th Battalion, was marked with a black heart superimposed on a white diamond. A white numeral five was painted next to this symbol. These markings were applied to all of the tanks before the units were rail-loaded for the front.[37]

On Saturday the seventh, both battalions loaded one and a half companies on trains and dispatched them to the front. The 327th elements were bound for the Bois de la Hazelle near Bernecourt, while the 326th's tanks would detrain and assemble in the Bois de Bausinsard.[38] That same night, half of Chanoine's tanks arrived amid a good deal of confusion. Most of them were moved to concealed positions in the Bois de la Reine, but one small group was mistakenly dropped eight miles away and was unable to link up. Inspecting the French units the next day, Patton arranged to move their assembly area to another wood so that the rest of their tanks—as well as his own, which were en route—could be swiftly unloaded and moved into concealed positions.[39]

The remaining American tank companies were loaded on trains at Bourg on 8 September. The first section, meanwhile, arrived in the forward area the next day to find the rail situation still in a state of flux despite Patton's efforts to get things sorted out. Trains loaded with heavy artillery pieces blocked the tracks, which were poorly constructed and in some cases crumpled under their weight. This and other problems led to delays and sidetracking of trains, further disrupting schedules. In some instances, units were required to detrain at other locations, requiring longer moves to assembly areas and causing traffic congestion.[40]

Patton, still worried about coordination with the 42d Division, met with Brig. Gen. Michael J. Lenihan, the assistant division commander, on Sunday the eighth. Patton was particularly concerned about the use of smoke and asked that provisions for the employment of smoke shells be incorporated in the division's fire-support plan. Lenihan referred Patton to the division's operations officer, Maj. Grayson M. P. Murphy, a West Point graduate who had resigned his commission before the war to become a banker, was called to organize the American Red Cross service in France when the United States entered the war, and was given the 42d Division assignment after he applied to return to active duty.[41]

Unfortunately, Murphy informed Patton, the request was too late. Patton noted in his diary that night that Murphy

> told me he could not put smoke in plan as stencil was already cut. The biggest fool remark I ever heard showing just what an S.O.B. the late chief of the Red Cross is. Told Col. [Stuart] Heintzelman [the 42d Division's chief of staff] of remark & said that if tanks fail in 42 Div it will be . . . Murphy's fault.[42]

In fairness to Major Murphy, it should be noted that the request for smoke was later incorporated into the corps artillery plan. The suggestion to get it into the division fire plan was just one of many that Patton made during the division's preparation for the battle. The pressure of getting the complicated and lengthy division operations plan and all of its supporting annexes typed, mimeographed, and distributed to higher, subordinate, and adjacent units prior to the deadline might easily have prompted Murphy's curt response to the nettlesome Patton, whose frustration was also understandable. His carefully thought-out plan for operations in the V Corps area had been discarded, and he was forced to go through the entire painstaking process again. In addition, rain had begun to fall on 7 September, and Patton was concerned that his hopes for battlefield glory for himself and the tanks might be swallowed in the quagmire he feared the Woëvre plain would become.[43]

Meeting with his battalion commanders on 9 September, Patton issued his operations order, which detailed the missions of his subordinate units and how they should conduct their attacks. The rain continued.

On the tenth, Lt. Col. Henry E. Mitchell, who previously had commanded the 301st Heavy Tank Battalion in England, and Major Viner arrived to observe the operation, which was set to kick off at 5 A.M. on Thursday the twelfth. Some elements of the 326th Battalion that had arrived the day before still had not detrained because the French had piled ammunition on the tracks. Other problems plagued the remaining companies arriving with the second section of trains the night of 10–11 September. French railroad officials either could not or would not stop one train where the tanks could get off, so someone threatened the engineer, forcing him to halt. Viner, working throughout the day on the eleventh, was able to get all but Company C of the 327th Battalion off the trains.[44] And still the rain fell.

Rockenbach received word from the commander of the French Tank Corps on the morning of the eleventh that the new Renaults his units had received should not be employed before being broken in for at least twelve days. Another message came in later that afternoon saying that, because of the rain, the tanks would not be able to operate as planned. Rockenbach, anxious to get his tanks into the fight, replied that "as long as we did not have over two inches of mud, it would act as a lubricant, and we would operate."[45]

The tanks moved to their final positions on the night of 11 September, the two French groupes and the 327th Battalion to the eastern edge of the Bois de Remierres, and the 326th Battalion to just north of Xivray-Marvoisin. Patton took this opportunity to issue his final instructions before the attack. Among other things, he specified that he wanted all cooking completed before dark and that units were to maintain strict light discipline (meaning no light of any type was to be permitted after sunset). He also warned the men to beware of the water in streams, as it was highly possible it might be contaminated by gas. Finally, he wanted the men to know that:

> From a tactical point of view the present operation is easy. A complete success insures the future of the Tank Corps in which all have shown by their long and cheerful work that they are fully interested. . . .

No tank is to be surrendered or abandoned to the enemy. If you are left alone in the midst of the enemy keep shooting. If your gun is disabled use your pistols and squash the enemy with your tracks. . . . If your motor is stalled and your gun broken still the infantry cannot hurt. You hang on [and] help will come. In any case remember you are the first American tanks. You must establish the fact that AMERICAN TANKS DO NOT SURRENDER. . . .

As long as one tank is able to move it must go forward. Its presence will save the lives of hundreds of infantry and kill many Germans.

Finally This is our BIG CHANCE; WHAT WE HAVE WORKED FOR . . . MAKE IT WORTH WHILE [emphasis in original].[46]

By 1 A.M. on 12 September all of the 1st Tank Brigade's elements were in position for the attack with the exception of Compton's Company C, which was still detraining in the vicinity of Ansauville.[47]

Pfc Melvin Winget, serving as a runner for Capt. Courtney H. Barnard, the Company C commander, recalled that "we had quite a time unloading our tanks from the train and getting them into position. I don't know how many miles [we had to move them] in the dark and rain, but we got there in time to shove off at about 5 A.M."[48]

Getting to the line of departure proved to be extremely difficult for the men of Brett's 326th Battalion. The concern Patton and Rockenbach had expressed for the effects of rain on the ground in the sector was well founded. Brett recalled later that the mud was knee deep, severely hampering movement.[49]

Initial efforts to maintain noise and light discipline while moving forward in the dark were effective. Sgt. D. M. Taylor, one of Brett's runners, recalled that as the battalion began its move to the line of departure, "it was silent and secret as any inner chamber session of the Masons or XP's," as soldiers spoke only in whispers and carefully placed their feet to avoid making loud splashes in the muck.[50] But confusion soon reigned. Despite the careful work of the battalion reconnaissance officer, Lieutenant Heilner, who had marked the route with white tape the night before, men and tanks soon became lost in the inky darkness. Heilner, near the head of the column, was unable to find the tape even after sending several men in different directions to search for it. Finally, he located one of the engineers he had asked to be posted along the route at 100-meter intervals, and was able to mark the route for the tanks by posting soldiers every ten paces as guides.[51] Discipline broke down in the midst of this confusion, however, and men were soon shouting instructions back and forth, lighting cigarettes, and using flashlights to guide tanks into position or firing pistols to get attention.[52] It took more than three hours, but by midnight the battalion had completed the half-mile move from its assembly area to the line of departure.

Following a four-hour artillery preparation, both battalions and the two French groupes moved out as planned, the 326th Battalion's tanks leading the 1st Infantry Division's doughboys on the left, and the other tanks following the lead elements of the 42d Infantry Division's 83d Brigade on the right.[53] Patton, perched on a hill in front of the main line, was positioned so he could watch the advance of his entire brigade. He recalled that it was dark and there was heavy rain and wind. Although visibility was poor because of the fog, he was able to report telephonically to IV Corps at 6:10 A.M. that Brett's tanks were passing Xivray and that there was little German response to the American shelling.[54]

It was not long, however, before Patton's vantage point became a source of irritation:

> I could see them coming along and getting stuck in the trenches. It was a most irritating sight. At 7 o'clock I moved forward 2 miles . . . and passed some dead and wounded. I saw one fellow in a shell hole holding his rifle and sitting down. I thought he was hiding and went to cuss him out, he had a bullet [hole] over his right eye and was dead. As my telephone wire ran out at this point I left the adjutant there and went forward with a lieutenant and 4 runners to find the tanks, the whole country was alive with them crawling over trenches and into woods.[55]

Patton's next report to IV Corps, dispatched by motorcycle messenger at 7:20 A.M., said that at least sixteen tanks were engaged in heavy fighting and added that the artillery smoke screen fired by the division and corps was extremely effective. Moving to a hill some 800 meters northwest of Seicheprey, Patton reported an hour later that his tanks were preceding the infantry in both divisions' sectors and that from his vantage point only five tanks appeared to be out of action from undetermined causes.[56]

Had Patton's view of the battlefield been better, he might have lost his temper. Compton recalled later that by 9 A.M. he had only twenty-five tanks actively engaging the enemy. Another ten had been disabled, ten more were down at or near the line of departure, and three were preparing to go into action. Compton was also employing seven tanks in a supply role and had been tasked with providing the sixteen tanks for brigade reserve. Most of his disabled tanks were from the late-arriving Company C. As quickly as possible, seven were repaired and refueled and accompanied him as the battalion reserve.[57]

Capt. Dean M. Gilfillan, commander of the 327th's Company A, led one of his platoons through a hail of fire in an attack on the southern edge of the Bois de la Sonnard, knocking out a number of machine-gun nests. But that was the only real resistance put up by the Germans in Compton's sector during the battalion's first four hours of combat. At 9:15, Compton sent a report to Patton that the already difficult terrain had become virtually impassable because of the five days of rain, and his tanks were being delayed.[58]

Unable to see tanks from any of his subordinate units at this point, Patton, accompanied by another officer (presumably Lieutenant Knowles) and three runners, set off in search of Compton. As Patton moved forward he noted that there were few dead Germans in the trenches, indicating only slight resistance.[59]

While the fight was not shaping up as a true test of the combat power of the tanks, the machines and men were nevertheless being taxed to the utmost. The Renaults, designed to cross six-foot trenches in dry weather, were being forced by their crews to negotiate line after line of trenches that were eight feet deep and ten to fourteen feet wide "in horrible mud."[60] Many did not make it. First Lt. Don C. Wilson of the 1st Brigade's 321st Repair and Salvage Company recalled later that mud was the real enemy—not the kind that somehow finds its way into radiators or carburetors, "but sticky, soggy, awful mud in which the tanks wallowed belly deep." Wilson spent the entire operation moving around the battlefield extricating one tank after another. The work was not as monotonous as it might sound. He soon discovered that each tank's predicament was unique and required him to employ different tactics to recover it.[61]

As Patton moved forward he encountered sporadic shelling and admitted to a desire to duck but said he soon saw it would do no good to attempt to avoid fate. "Besides, I was the only officer around who had left on his shoulder straps and I had to live up to them. It was much easier than you would think and the feeling, foolish probably, of being admired by the men lying down is a great stimulus."[62]

On the left flank of the IV Corps sector, Major Brett and his 326th Battalion encountered similar tough going. One of Brett's platoons, attacking the Salient du Harem east of the Rupt de Mad, managed to cut the wire in front of the trenches. The remaining tanks, operating west of the river, cleared the sector in advance of the infantry between Reichecourt and the river in the direction of Lahayville, then assaulted machine-gun positions in the Bois de Rate.[63]

It should be noted that battles are rarely the simple, straightforward actions official reports make them out to be. Such was the case with the 326th Battalion's activities that Thursday morning. From his vantage point as a runner for the battalion commander, Sergeant Taylor observed that the fighting quickly deteriorated into individual skirmishes, adding that "it looked as if there were at least six different wars going on instead of one."[64]

Brett lost his first tank in a ditch just after crossing the line of departure and, pigeon basket in hand, commandeered the nearest vehicle. Within minutes, this tank too was ditched. He took over a third vehicle but was forced to abandon it when it became entangled in rubble as the battalion passed through a village. "Thoroughly disgusted" with this turn of events, Brett ignored Patton's order to remain 1,000 meters behind his lead elements and set off on foot to direct the attack.[65] Other officers and men soon followed suit. Brett recalled seeing his three company commanders, Captains Weed, Semmes, and English, standing on trench parapets in the face of heavy fire, directing their units over the difficult terrain.[66]

Semmes had lost his tank while trying to cross the Rupt de Mad. When he escaped from the submerged vehicle, he realized that his driver was trapped inside and, in the face of heavy machine-gun fire and at the risk of his own life, dove back under to rescue him.[67]

Second Lt. Julian K. Morrison, one of Semmes's platoon leaders, came upon a German machine-gun nest as he led his tanks forward on foot. Struck by two bullets in the right hand, he continued forward and, armed only with his .45-caliber pistol, captured the crew.[68]

Typical of the experiences of Brett's leaders was that of 2d Lt. Edwin A. McClure, a platoon leader in Company A, who found himself

guiding tank on foot across the trenches and pulling all of the tanks of the platoon out of the positions in which they had become stuck; opening numerous lanes through the enemy wire and thru the dense woods for the infantry; bridging [an] extra wide and deep trench by collecting duckboards from its bottom; getting tank tangled in overhead wire in the woods and having a bad three minutes breaking loose while our barrage, falling short [burst] around me; putting flight about a dozen machine-gun crews, causing

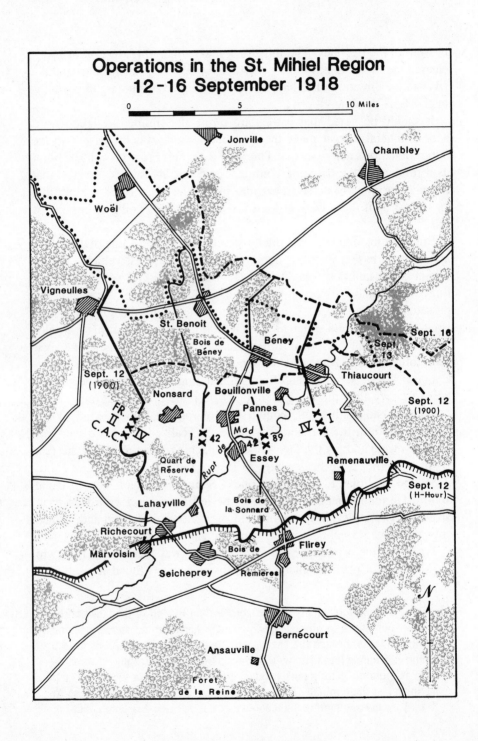

Operations in the St. Mihiel Region
12-16 September 1918

them to abandon material without inflicting casualties; getting in a smoke barrage while changing from one sector to another and being nearly blinded by burning ashes coming through the [tank's vision] slits; leading tank out of the blinding choking smoke wave, guided by the sound of clanking tanks, finally [running] out of gas and going into Nonsard [the battalion's objective] on foot with the tanks as an emergency commander or driver, taking prisoners with the [first] patrol to enter the town, exploring along a house on the outskirts and taking five Boche [Germans] from its basement, driving them out by using beer bottles as grenades.[69]

As the battalion approached Nonsard, Brett linked up with eight tanks from the brigade reserve that had joined the fight and, after ordering a lieutenant to drive for him, led the force in advance of the infantry to the objective. Sergeant Taylor watched from a rise overlooking the town as Brett silenced the only apparent opposition, a machine-gunner in a church steeple, with his tank's 37mm gun. As the first wave of infantry streamed past his position, Taylor noted that the doughboys seemed unconcerned. The first man to pass him had his rifle slung over his shoulder and was nonchalantly rolling a cigarette. The second was carrying a head of cabbage in one hand and busily feeding himself hardtack with the other.[70]

In the 42d Division sector, Major Chanoine and his Schneider tanks were slowed to a crawl by the wide, muddy trenches. Still, he managed to keep them moving steadily, following closely on the heels of the infantry until he was ordered to halt just short of Essey.

Patton, still in search of Compton's battalion on the right flank, encountered Chanoine and stopped to talk briefly with the French officer as he supervised work on a stalled vehicle. As Patton walked away, a 150mm German shell struck the tank amidships, wounding several men. One fragment struck Chanoine's helmet, knocking him unconscious. "But he was up in an instant and exclaiming: 'All right now we wont have to fool with the old cow,'" and waving the rest of his tanks forward.[71]

Moving on toward Essey, Patton encountered the lead elements of the 83d Brigade. The doughboys were lying down, but as there was only scattered artillery fire, Patton continued on to a small hill where he found the brigade commander, Brig. Gen. Douglas MacArthur, standing and watching the German retreat:

> I joined him and the creeping barrage came along toward us, but it was very thin and not dangerous. I think each one wanted to leave but each hated to say so, so we let it come over us.

We stood and talked but neither was much interested in what the other said as we could not get our minds off the shells.[72]

By this time, elements of the 327th Battalion were also in the vicinity of Essey. Their approach had not been entirely without incident. Compton reported later that in one instance "twenty Boche came up to the tank of Corporal Pattison and surrendered."[73]

The move across the rough terrain in the salient kept 1st Lt. Tom W. Saul, the 327th's repair officer, gainfully employed as he directed numerous recovery operations aimed at repairing mechanical breakdowns and pulling ditched tanks out of the mud. He also gave directions to engineer officers tasked with making lanes for the tanks through the trench system and personally led a number of tanks through the maze in the face of heavy German fire.[74]

Outside Essey, Patton was approached by five 327th Battalion tanks. When the lieutenant in charge asked for instructions, Patton told him to continue on through the town. Patton watched as the lieutenant led his platoon forward, only to turn back when "some damned Frenchman at the bridge told them to go back as there were too many shells [falling] in the town." This infuriated Patton, who led the tanks into Essey on foot.[75]

As the doughboys and tanks moved through the town, a number of Germans came up from their dugouts and surrendered to MacArthur and Patton, who were following their units.[76] Pfc Newman Chilcress, a runner in the 327th, also managed to capture a machine gun and its four-man crew.[77]

The attack continued smoothly toward Pannes with MacArthur's blessing, although the road "was rather a mess." A German artillery battery apparently had fallen prey to an American barrage that left the mangled corpses of men and horses strewn along it.[78]

Patton, Knowles, and a sergeant named Graham who was serving as a runner mounted the lead tank and preceded the infantry into Pannes. Knowles and Graham watched one side of the road while Patton kept an eye on the other. As they approached a crossroads in the center of town, Knowles and Graham leaped from the tank, pistols drawn, and chased a fleeing German into a house. Inside they found an entire platoon. They took a total of thirty prisoners, who were later turned over to the infantry.[79]

On the far side of town, Patton continued to ride the back deck of the

Renault as it headed blithely off in the direction of Beney. Several hundred yards outside Pannes a burst of machine-gun fire struck the opposite side of the tank, showering Patton with paint chips and sending him in search of a shell hole for safety. The hole in which he took cover was a shallow one, and every time he moved "the boshe shot at me" and "bullets nocked all of the front edge [of the hole] in on me." Meanwhile:

> The tank had not seen me get off and was going on. The infantry was about 200 [yards] back of me and did not advance. One runner on my right got hit.
> I was on the point of getting scared as I was . . . all alone in the field. If I went back the infantry would think I was running and there was no reason [for me] to go forward alone. [80]

Faced with the dilemma of how to get back to the infantry lines without appearing to retreat, Patton elected to move obliquely from shell hole to shell hole, trying to elude the German machine-gunners. When he finally reached the 167th Infantry's forward elements (the tank all the while continuing its advance), Patton asked the commander to move forward in support of his vehicle. He was told that the infantry's sister battalion on the right had failed to move up, and the commander had been instructed to hold at Pannes. Patton then asked for a runner to go forward to halt the tank and bring it back, to which the "heroic" infantry commander is said to have replied, "Hell no. It ain't my tank."[81]

Not wanting the tank to proceed into Beney alone, Patton set out on foot to bring it back. The idea of rescuing the tank and its crew fixed firmly in his mind, Patton recalled later that he "was not the least scared." He did admit to having "run like H——," however.[82]

At about 12:30, the remaining four tanks from the platoon arrived on the scene. Patton, finding no officer in charge, gave the platoon to Lieutenant Knowles and coordinated plans with the infantry commander, who this time agreed to support an attack on the village.

The second attack went much more smoothly. By this time most of the German machine-gunners were either dead or had been scared off by the tanks. But as the lead elements drew to within 400 yards of the town, the infantry suddenly veered off to the right in the direction of the Bois de Thiacourt, which had been assigned to them as a new objective. Patton told Knowles to have the tanks change direction and lead the infantry into the woods. Once the infantry halted, the tanks withdrew toward Pannes and at about 2 P.M. were reinforced by two more tanks.

Two additional tanks, recently refueled, arrived from the 327th Battalion's reserve. Because of the mud and the number and width of the trenches, the tanks were burning fuel about three times faster than expected. Although a small amount had been towed forward on sleds by Compton's supply tanks, by 3 P.M. the fuel situation was critical. Efforts to move fuel by truck to Essey on the Flirey-Essey road were frustrated by military police, who halted three of the brigade's fuel trucks at Flirey and refused to allow them forward until after 2 P.M. the next day, severely curtailing operations on the morning of the thirteenth.[83]

With the arrival of the two reserve tanks, the 327th launched its final assault on Beney, this time supported by a platoon from the 167th Infantry. They cleared the town and continued to the Bois de Beney to the northeast, encountering only light opposition. During the operation the tanks captured sixteen machine guns and a battery of four 77mm guns in the town.[84]

Satisfied that the enemy had withdrawn from the area, the attack force pulled back to where the main body of the 167th Infantry was digging in just south of Beney. The tanks continued to the rallying point in Pannes. One of the tanks was disabled in a field south of Beney and became a target for German artillery, which shelled it intermittently throughout the night.

Compton wrote later that his battalion had operated with great success on the twelfth, advancing fifteen kilometers at the farthest point and materially assisting the infantry in capturing "great quantities of machine-guns, field pieces of large and small caliber, and immense quantities of stores and supplies in towns."[85]

Patton, also satisfied with the results of the attack on Beney, submitted a report on the encounter to IV Corps at about 3 P.M., then set off in search of Brett and the 326th Battalion in the 1st Division sector. Tired and hungry, Patton walked the three miles from Beney to Nonsard, where he found Brett, slightly wounded in the nose, sobbing with frustration: Only twenty-five tanks had made it with him to the objective, and, worst of all, they had no fuel left, so he was unable to continue the attack. That was the kind of attitude Patton could appreciate, and he did all he could to console Brett before heading to the rear in search of gasoline.[86]

Two of Compton's supply tanks were able to drag some gasoline on sleds from Bernecourt to Pannes during the night, but that was the only fuel to reach the front before 10 A.M. on Friday the thirteenth.

At about 6:30 that morning fifteen tanks, including the tanks of the 327th Battalion's three company commanders, reported to Brigadier

General MacArthur in the vicinity of St. Benoit. His brigade's mission accomplished, he directed the tank officers to move to a wooded area in the rear where he could easily call for them if needed and told them to camouflage their positions to conceal the tanks from low-flying German aircraft.

Late in the morning, 1st Lt. Gus Struyk, the 327th's supply officer, arrived in Pannes with four trucks loaded with rations, gasoline, oil, and ammunition. Patton ordered that the supplies be divided between Brett and Compton, and two trucks were dispatched to Nonsard with an equal share for the 326th Battalion.[87] Compton was able to top off another twenty tanks by noon and sent them forward to St. Benoit, giving Mac-Arthur a reserve of thirty-five tanks. At about 2 P.M. Struyk's remaining supply trucks reached Brett, who was able to refuel fifty tanks and get them rolling to Vigneulles, where they assembled about midnight.[88]

Compton received an order from Patton at 11:30 P.M. on 13 September to "go to the assistance of the infantry against a boche counter attack." But the attack never materialized, and the 327th Battalion's active participation in the offensive came to an end. Compton's tankers spent most of the fourteenth performing minor repairs and tuning up the tanks in preparation for the move to the Meuse-Argonne sector. Late that afternoon the 327th was ordered back to its original assembly area to prepare for rail-loading, and movement back to the Bois de la Hazelle began at 7 P.M. With few exceptions, the battalion's tanks rallied at their destination by 1 A.M. on Sunday the fifteenth.[89]

Patton met with Brett, who had been unable to make contact with any of the 1st Division's forward elements, early on the fourteenth. Impatient to move forward, and knowing that part of the American front lay along the Woël–St. Benoit road, Patton decided to send Brett through St. Maurice and on to Woël in search of the infantry.[90]

Brett moved out with his entire battalion at 6 A.M., and at 6:45, Patton notified Captain Hebert back at the brigade command post to pass the word to Rockenbach and IV Corps that Brett was headed north with fifty-one tanks. Patton added that he expected to get back to the command post "as soon as possible" and told the adjutant to wait there for him. He then hustled off in an effort to catch up with Brett.[91]

By 9 A.M. Patton had caught up with the battalion, which was halted about two kilometers west of Woël. Captain Weed recalled that Patton seemed unconcerned that the battalion still had been unable to locate any American infantry units. After "consulting his speeches to the Staff College," Weed later wrote, Patton pointed out to Brett and his officers that, lacking infantry support, they "were therefore 'a cavalry patrol.' Just what the Boche thought, no one knows, but he was mighty inquisitive and we had several callers in the shape of avions" who fired on Brett's column.[92]

As Patton discussed the situation with Brett, he recognized Brig. Gen. Dennis E. Nolan, Pershing's G-2 (intelligence officer) throughout the war, passing by in an automobile and flagged him down. When Patton informed Nolan of his destination and that they were looking both "for a fight" and for the 1st Division, Nolan told him Woël had been evacuated by the Germans and was being held by a platoon of French infantry.[93]

Patton ordered Brett to hide his tanks in the shrubbery along the side of the road and sent a message to Rockenbach and the IV Corps and 1st Division commanders asking for further instructions. While awaiting this guidance he also sent four officers, using captured German horses as mounts, on a patrol of the woods to the south in search of American infantry. They found no one.

During the two hours this patrol was out, a small convoy of supply trucks arrived. The trucks had been attacked by a German airplane, and one of the soldiers on board was wounded by a bomb fragment that went through his right arm. Brett took the opportunity to have his tanks topped off with fuel and fed his troops. By noon, still without any word from higher headquarters, Patton and Brett decided to send a patrol of three tanks and five dismounted soldiers to Woël, then down the road an additional two kilometers toward St. Benoit. At about 1:30, Lieutenant McClure, who was leading the patrol, reported the town was indeed clear of enemy troops and that he was returning. Thirty minutes later the patrol was attacked by a force estimated to be at least a battalion of infantry accompanied by a battery of 77mm guns. McClure sent a runner back to the battalion with word that he was attacking. Patton immediately ordered a platoon of five tanks led by 2d Lt. Gordon M. Grant, also from Company A, to go to his assistance. They linked up with McClure's three tanks at 2:30 and, still without infantry support, drove the Germans about six kilometers to the outskirts of Jonville. During the running battle the tankers killed or put to flight at least a dozen machine-gun crews and captured four 77mm cannon. The patrol was shelled while the men were attempting to attach the artillery pieces to the backs of their tanks, and both McClure and Grant and four of their soldiers were wounded by shrapnel. McClure ordered the breechblocks removed and the guns abandoned. Two of the tanks, which had been mechanically disabled, were coupled to other tanks and towed back to the battalion. The patrol cleared out just in time as the Germans hammered its previous position with a barrage of 150mm artillery fire.[94]

Artillery fire had also begun to register on the rest of Brett's battalion, and Patton decided to pull the force back to St. Maurice.

At 9 P.M. on 14 September the Tank Corps's participation in the St. Mihiel offensive officially came to a close, when all tanks were ordered to concentrate in the vicinity of the Bois de la Hazelle to prepare for the rail move to the Meuse-Argonne sector. Moving under cover of darkness, all of the French and American tanks were consolidated in their original assembly areas by the night of the sixteenth. Vehicle losses during the withdrawal were three French and two American tanks, all damaged by direct artillery hits.[95]

The Tank Corps experienced relatively minor casualties during the battle, most of them on the first day of the operation. In Patton's two battalions four officers were wounded (one while in a tank). Five enlisted men were killed, and fifteen were wounded (three while in tanks).[96] Casualties in the attached French units were higher: five officers wounded (none in tanks), six enlisted men killed (one in a tank), and twenty-three enlisted men wounded (none in tanks).[97]

By 16 September Patton's brigade was down to 131 operational tanks. Three had been knocked out by direct artillery hits, and forty more were still mired in the trenches or out of action with mechanical problems.[98]

While the Tank Corps's participation in the St. Mihiel offensive was of short duration and not especially violent, it gave the tankers a chance to "get rid of nervousness" and allowed them to try out maneuver tactics on the battlefield so they could "settle down to business in subsequent operations."[99]

Both battalions had tried to employ tactics worked out during their training at Bourg. They moved out from the line of departure in a formation that allowed them a depth of at least three lines of tanks. This was accomplished by deploying two companies in the lead echelon, each company leading with two platoons on line and retaining the third platoon in support. The third company in each battalion was placed in battalion reserve.[100] Unfortunately, the speed of the German withdrawal, coupled with the difficult terrain and mechanical difficulties, forced the tankers to abandon their formations and operate alone or in small groups. Nevertheless, their mere presence on the battlefield "gave great moral support to the infantry and demoralized the Boche." The latter effect must have been somewhat difficult to assess, however, because, as Patton observed in his after-action report, the Germans failed to put up a serious fight, and thus "the full value of tanks in this operation was not possible to demonstrate."[101]

But the battle did demonstrate the effectiveness of Patton's training methods and ability to motivate soldiers. Lt. Julian K. Morrison said every officer in the Tank Corps knew, after listening to Patton's lectures, that a tank officer was "meant to die." Morrison said Patton's

> favorite message to his officers was "Go forward, go forward. If your tank breaks down go forward with the Infantry. There will be no excuse for your failure in this, and if I find any tank officer behind the front line of infantry I will [probably 'shoot him']."[102]

Morrison said that the officers passed this message on to their men, so it was not unusual to see tankers stretched from one to seven kilometers in front of the infantry. Morrison also recalled, perhaps with some exaggeration, that everyone in the Tank Corps at St. Mihiel was caught up with Patton's fighting spirit, with the result that cooks, company clerks, first sergeants, mechanics, and runners joined in the fray.[103]

Probably more significant than anything else that came from the Tank Corps's participation in the St. Mihiel offensive was the impact of the small action near Jonville on Patton's thinking. Until that time, Patton had adhered to the idea that tanks were strictly an infantry support asset. McClure's battle provided him with a vision of what more mechanically advanced tanks might be able to accomplish on future battlefields operating as an independent combat arm.[104] Patton also exhibited that rare ability to adjust quickly to a rapidly changing situation on the battlefield. This trait would later become a hallmark of his World War II operations.

The personal presence of Patton, Brett, and Compton on the battlefield, while a motivating factor for their men, served only to incense Rockenbach. After the battle he chastised Patton for failing to maintain closer contact with his headquarters. Rockenbach assured Patton that brigade and battalion commanders had no business charging around on the battlefield. He told Patton that in the future he expected him to remain at his command post and battalion commanders to remain with their reserve elements. Only company commanders and platoon leaders should be with the forward elements.

The fact that so many of the brigade's leaders could be found on foot leading their tanks into battle was attributable less to a desire to inspire the troops than to the poor means of communications available to them. Captain Semmes later recalled that word of mouth was the only reliable means of conveying orders in combat. Attempts to use flags poked through a hole in the turret usually resulted in the flag being shot off as fast as it appeared, and the "would-be signaler invariably ended up with splinters in his hand."[105]

The use of homing pigeons also proved fruitless. Semmes recalled that each command tank was supplied with a wicker basket containing two pigeons. The baskets were about six inches high and more than a foot square, so it was necessary to place them on the turret floor. Given the crowded conditions in the turret, where the gunner found himself competing for space with guns and ammunition, it was only natural that the baskets wound up as footrests. "Unfortunately," Semmes wrote, "the designers had not planned for such an eventuality, so nearly every pigeon basket in the first five minutes of action was flattened out, and, of course, the pigeons were flattened with the baskets. Thus, no pigeon messages got back from the tanks to headquarters."[106]

Rockenbach also told Patton that, while he appreciated the Tank Corps's esprit de corps and the men's personal valor, he wanted them to know "that they are fighting [in] tanks, they are not Infantry, and any man who abandons his Tank will in the future be tried [by court-martial]."[107]

While Patton dutifully passed on Rockenbach's message to his subordinate officers and the troops, he failed to take the instructions to heart. In a letter to his wife written after the battle he noted that "Gen. R. gave me hell for going up [with the forward elements] but it had to be done. At least I will not sit in a dug out and have my men out in the fighting."[108] Patton's proclivity for being in the thick of action would manifest itself again during the Meuse-Argonne campaign and throughout the remainder of his career.

Notes

1. Russell F. Weigley, *The American Way of War: A History of United States Military Strategy and Policy* (Bloomington, Ind.: Indiana University Press, 1977), 202–3.
2. Edward M. Coffman, *The War to End All Wars: The American Military Experience in World War I* (New York: Oxford University Press, 1968), 263.
3. As quoted ibid., 169.
4. Ibid., 169–71.
5. Ibid., 171. World War I U.S. Army square divisions each had two brigades. The brigades were numbered consecutively throughout the Army, with the 1st and 2d Brigades in the 1st Division, and so on. The 4th Brigade of the 2d Division went on to become known as the Marine Brigade.
6. Ibid., 263.
7. Allan R. Millett and Peter Maslowski, *For the Common Defense: A Military History of the United States of America* (New York: The Free Press, A Division of Macmillan, Inc., 1984), 354.
8. As quoted in Coffman, *War to End All Wars*, 262.
9. Rockenbach, "Tank Corps Operations," 7.
10. Ibid., 8.
11. Ibid.
12. Ibid.
13. As quoted in Blumenson, *Patton Papers*, 569.
14. As quoted ibid., 570.
15. Ibid.
16. Rockenbach, "Tank Corps Operations," 8.
17. Ibid., 8–9.
18. Ibid., 9.
19. Blumenson, *Patton Papers*, 571.
20. Patton diary entry, 28 Aug. 1918, as quoted ibid., 572.
21. Ibid., 572–73.
22. Patton diary entry, 30 Aug. 1918, as quoted ibid., 573.
23. George S. Patton, Jr., to Beatrice Ayer Patton, 31 Aug. 1918, as quoted ibid.
24. Rockenbach, "Tank Corps Operations," 8–9.
25. Col. George S. Patton, Jr., "Tanks, Tankers, and Tactics, Chapter 4," Patton Writings, Box 59, Tanks (Reports) 1909–1944, Patton Collection.
26. Ibid.
27. Rockenbach, "Tank Corps Operations," 10.

28. Lt. Col. Emile Wahl, "Operations of the French Tank Corps with the First American Army," included as Appendix 7 to Rockenbach, "Tank Corps Operations," 1.
29. Coffman, *War to End All Wars*, 273–75.
30. Ibid.
31. Blumenson, *Patton Papers*, 574.
32. Ibid.
33. Rockenbach, "Tank Corps Operations," 12.
34. Ibid., 3; Blumenson, *Patton Papers*, 575.
35. Blumenson, *Patton Papers*, 577.
36. Capt. Ranulf Compton, "War Diary of the 345th Tank Battalion," World War I Survey Collection, Archives, USAMHI.
37. Maj. Gen. William R. Kraft, Jr. (U.S. Army Ret.), "The Saga of the Five of Hearts," *Armor* 97 (July–Aug. 1988): 35.
38. Compton, "345th Battalion War Diary."
39. Blumenson, *Patton Papers*, 578.
40. Ibid., 579–80.
41. Ibid., 578.
42. Patton diary entry, 8 Sept. 1918, as quoted ibid., 578–79.
43. Ibid., 579.
44. Compton, "345th Battalion War Diary."
45. Rockenbach, "Tank Corps Operations," 10.
46. Lt. Col. George S. Patton, Jr., "Special Instructions for the 326th Bn., and 327th Bn.," Patton Chronological Files, Box 10, 8 Sept. 1918, Patton Collection.
47. Col. George S. Patton, Jr., "Operations of the 304th Tank Brigade, September 12th to 15th, 1918, St. Mihiel Salient," 12 Nov. 1918, included as Appendix 4 to Rockenbach, "Tank Corps Operations," 1; Compton, "345th Battalion War Diary."
48. Questionnaire, Pfc Melvin Winget, Company C, 345th Tank Battalion, World War I Survey Collection, Archives, USAMHI.
49. Maj. Sereno E. Brett, "Report of Major S. E. Brett," 14 Dec. 1918, Patton Military Papers, Box 47, Personal Experience Reports of Tank Operations—1918, Patton Collection.
50. 2d Lt. D. M. Taylor, "My Horrible Experiences," undated, ibid.
51. 2d Lt. George B. Heilner, "Personal Experience Report," 14 Dec. 1918, ibid.
52. Taylor, "My Horrible Experiences."
53. Some confusion exists over the designation of Patton's two light tank battalions during the St. Mihiel offensive. Word arrived in France in early September that two light tank battalions bearing the same numerical designations as Patton's were en route to the AEF from Camp Summerall, Pennsylvania. (See *Order of Battle*, 1544.) Because of this, Rockenbach's

headquarters issued General Order no. 16 on 12 September ordering the redesignation of the 326th and 327th Battalions to the 344th and 345th Battalions, respectively. The order also called for the 311th Tank Center to become the 302d Tank Center, and Patton's 316th Repair and Salvage Company to become the 321st R&S Company. (See Modern Military Records Division, Record Group 120, Entry 1296, NA.) I have elected to continue to refer to Patton's battalions by their original designations as these appear in most of the pertinent documents for St. Mihiel. By the beginning of the Meuse-Argonne campaign the new designations were consistently applied in the records.

54. Blumenson, *Patton Papers*, 584.
55. As quoted in Blumenson, *Patton Papers*, 584.
56. Ibid.
57. Compton, "345th Battalion War Diary."
58. Ibid.
59. Blumenson, *Patton Papers*, 585.
60. Patton, "Tanks, Tankers, and Tactics."
61. 1st Lt. Don C. Wilson, "Personal Experience Report," undated, Patton Military Papers, Box 47, Personal Experience Reports of Tank Operations—1918, Patton Collection.
62. As quoted in Blumenson, *Patton Papers*, 585.
63. Patton, "Appendix 4, Tank Corps Operations," 3.
64. Taylor, "My Horrible Experiences."
65. Ibid.
66. Brett, "Report of Major Brett."
67. General Order no. 24, General Headquarters—Tank Corps, American Expeditionary Forces, Subject: Award of Distinguished Service Cross, 17 Dec. 1918, Modern Military Records Division, Record Group 165, Entry 310, Box 446, File 66-64.2, NA.
68. Col. George S. Patton, Jr., "Citation of Members of First Brigade, Tank Corps, for Gallantry," 3 Nov. 1918, Patton Chronological Files, Box 11, 3-6 Nov. 1918, Patton Collection.
69. 1st Lt. Edwin A. McClure, "Personal Experience Report," undated, Patton Military Papers, Box 47, Personal Experience Reports of Tank Operations—1918, Patton Collection.
70. Taylor, "My Horrible Experiences."
71. Patton, "Tanks, Tankers, and Tactics."
72. As quoted in Blumenson, *Patton Papers*, 585.
73. Compton, "345th Battalion War Diary."
74. Ibid.; General Order no. 24, GHQ—Tank Corps, AEF.
75. As quoted in Blumenson, *Patton Papers*, 585.
76. Ibid., 586.

77. Compton, "345th Battalion War Diary."
78. Blumenson, *Patton Papers*, 586.
79. Ibid., 586–87; Patton, "Appendix 4, Tank Corps Operations," 3.
80. As quoted ibid., 587.
81. Ibid., 587–89.
82. As quoted ibid., 588.
83. Patton, "Appendix 4, Tank Corps Operations," 4; Compton, "345th Battalion War Diary."
84. Compton, "345th Battalion War Diary."
85. Ibid.
86. Blumenson, *Patton Papers*, 589–90.
87. Compton, "345th Battalion War Diary."
88. Ibid.; Patton, "Appendix 4, Tank Corps Operations," 5; Blumenson, *Patton Papers*, 593.
89. Compton, "345th Battalion War Diary."
90. Blumenson, *Patton Papers*, 594.
91. Ibid.
92. Weed, "Personal Experience Report."
93. Blumenson, *Patton Papers*, 594.
94. Ibid., 595; McClure, "Personal Experience Report."
95. Patton, "Appendix 4, Tank Corps Operations," 5.
96. Rockenbach, "Tank Corps Operations," 14.
97. Ibid., "Appendix 5," 11.
98. Ibid., 14.
99. Ibid.
100. Patton, "Appendix 4, Tank Corps Operations," 4.
101. Ibid., 6.
102. As quoted in Blumenson, *Patton Papers*, 599.
103. Ibid.
104. Col. George S. Patton, Jr., to Capt. Shipley Thomas, 3 July 1919, Viner Collection no. 33, World War I Survey Collection, Archives, USAMHI.
105. Semmes, *Portrait of Patton*, 57.
106. Ibid.
107. Brig. Gen. Samuel D. Rockenbach to commanding officers of American Tank Brigades, Subject: Notes on American Tanks, 14 Sept. 1918, Patton Chronological Files, Box 10, 13–16 Sept. 1918, Patton Collection.
108. George S. Patton, Jr., to Beatrice Ayer Patton, 16 Sept. 1918, as quoted in Blumenson, *Patton Papers*, 597.

CHAPTER VII

The Meuse-Argonne, Phase One: Into the Valley of Death

W hile Patton's tankers worked to ready their vehicles for the rail move to the new front, First Army planners were busy mapping out a campaign that would prove to be the most ambitious military effort mounted by American forces up to that point in their history. More than a million Americans participated in the offensive, most of them without prior combat experience.

One of the problems Pershing brought on himself by accepting the St. Mihiel mission was the need to give his battle-tested divisions a break after the fighting in that sector, which in turn necessitated feeding untried divisions into the line on the first day of the new offensive.

Another problem was transportation. But, thanks to the efforts of a young staff officer, Col. George C. Marshall, Jr., the First Army was able to move more than 600,000 troops into the sector, replacing more than 220,000 French soldiers who cleared out after the end of the St. Mihiel offensive on 16 September and before H hour for the Meuse-Argonne campaign on the twenty-sixth. Many of the units involved came from as far as sixty miles away, with the average combat unit traveling forty-eight miles to get into position. Marshall accomplished this feat by putting all motor traffic on one of the three roads leading from the St. Mihiel sector, and marching troops and horse-drawn traffic on the other two. While this meant splitting up some units, it made for efficient use of the roads.[1]

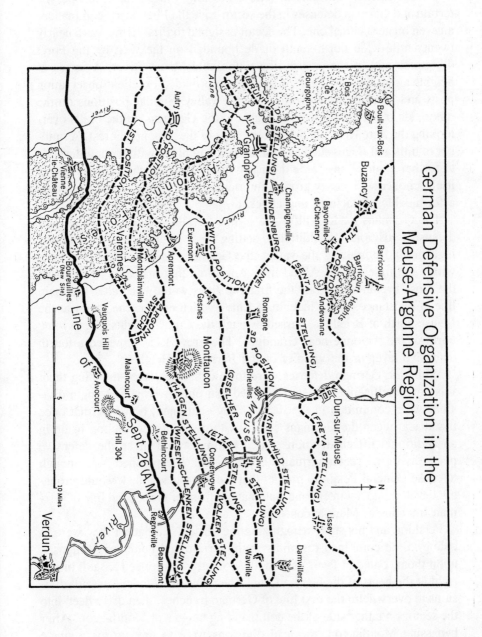

German Defensive Organization in the Meuse-Argonne Region

While Marshall was able to relieve Pershing's logistical headache, the terrain and German defenses in the sector gave the First Army commander an even bigger tactical one. The sector assigned to First Army was a nearly twenty-mile-wide north-south defile bounded on the west by the Forêt d'Argonne and in the east by the unfordable Meuse River. A series of heights along the east bank of the Meuse provided excellent observation posts and gun positions overlooking the valley. Similar positions on the other side of the sector were available to the Germans on the bluffs overlooking the Aire River on the eastern edge of the Argonne Forest. A number of hills and ridges such as Montfaucon combined with scattered woods to further channel any force daring to attack up the defile. The Argonne itself, choked with heavy growth and riddled with steep ravines, presented a formidable obstacle to any attacker.

The Germans made skillful use of this terrain when establishing their defensive positions. In addition to setting up a vicious cross fire from the high ground bordering the valley, the Germans had two defensive belts within ten miles of the Allied line. Any force that managed to negotiate the cross fire and penetrate the first two lines would encounter the main line of resistance – the Hindenburg Line – which entered the sector at the Meuse north of Brieulles, passed through the Cunel Heights, then swung north of the Argonne near Grandpré. The Freya Line provided a fourth defensive layer in the vicinity of the Barricourt Heights.[2]

While the German defenses were formidable, the force manning them was not. Intelligence officers estimated that only five understrength divisions were committed in the sector. They cautioned, however, that once the attack began the Germans could, within seventy-two hours, bring in an additional fifteen divisions. Such a force would make the defensive positions in the sector virtually impregnable. The prospect was enough to cause General Pétain to predict that the American attack would probably develop into a winter-long stalemate in front of the second line of German trenches at Montfaucon.[3]

Pershing and his staff, recognizing the virtual straitjacket they were in owing to the combined restraints of the terrain and boundaries imposed upon them, opted in favor of a swift, violent, two-pronged assault aimed right at the heart of the German line. The plan called for the assault divisions to overwhelm the first line of German trenches, then drive deep into the sector on either side of the commanding heights of Montfaucon. After bypassing Montfaucon, the lead divisions were to link up for a single thrust aimed at the Hindenburg Line in the vicinity of Romagne and Cunel some ten miles beyond the jump-off line. All of this was to be accomplished on the first day, although the hoped-for penetration of the Hindenburg Line was not expected until the morning of the second.

It was an audacious plan, but Pershing and his staff were gambling that the preponderance of American forces in the sector, coupled with the element of surprise, would give them the necessary edge. There appeared to be an excellent chance to catch the Germans off guard. It was only natural for them to expect the Americans to exploit their success in the St. Mihiel offensive by driving on toward Metz or Briey. To reinforce this perception, French troops were retained to help man the outpost line in the Meuse-Argonne sector until the eve of the attack, and American reconnaissance officers wore French helmets and overcoats whenever they visited the line. Everything possible was done to lull the Germans into believing that only war-weary Frenchmen manned the trenches.[4]

Pershing's deception plan also called for a demonstration by elements of Patton's tank brigade east of the Moselle River in the vicinity of Pont-à-Mousson. Major Viner, who in addition to commanding the 302d Tank Center at Bourg was also the Tank Corps's assistant chief of staff, was ordered to report to Col. Stuart Heintzelman, IV Corps chief of staff, on 18 September to coordinate the employment of fifteen tanks in a demonstration near Liverdun the night of 19–20 September.[5]

While Viner made his way to IV Corps headquarters, 1st Lt. Ernest A. Higgins took charge of a force of four other officers and sixty enlisted men manning fifteen tanks, two trucks, and a captured German rolling kitchen. With the exception of 1st Lt. Harry E. Gibbs, the brigade tactical officer, all of the men and equipment were drawn from Compton's 345th Battalion.[6]

Eighteen Pierce-Arrow trucks arrived in Bernecourt at 7 P.M. on the nineteenth to transport Higgins's force to the site of the demonstration. By this time it was apparent that the tanks would not reach the sector in time to execute the operation as originally planned. After the tankers loaded eleven tanks onto the trucks with much difficulty, word arrived that they should be unloaded, the trucks released, and the force prepared to rail-load at 8 A.M. the next day. The train, delayed because of difficulty in obtaining narrow-gauge cars, did not arrive until 3 P.M. on the twentieth. The loading proceeded smoothly until the last tank had one of its tracks torn off while being loaded sidewise onto a flatcar. Higgins instructed the two crewmen

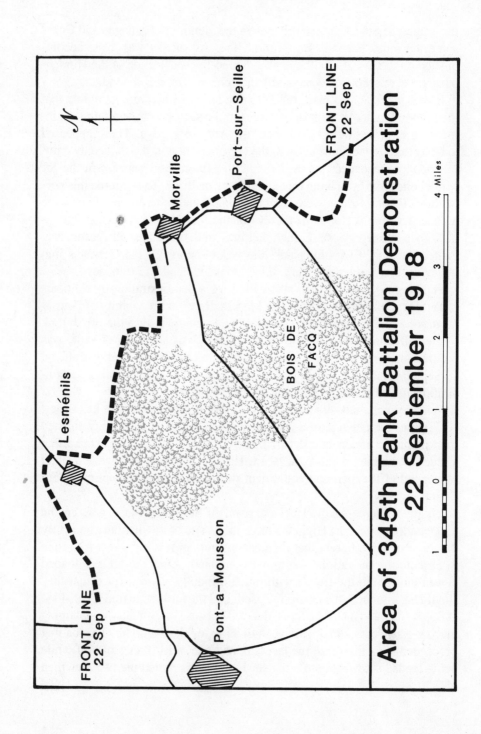

Lesménils

Morville

Port-sur-Seille

FRONT LINE
22 Sep

FRONT LINE
22 Sep

Pont-a-Mousson

BOIS DE
FACQ

N

**Area of 345th Tank Battalion Demonstration
22 September 1918**

1   0   1   2   3   4 Miles

to remain behind with their vehicle and join the rest of the battalion, which was entraining for the move to the Meuse-Argonne sector.

As the train was pulling out of Bernecourt, a runner arrived with orders for Higgins to report with his force to the Bois de Facq, located between Pont-à-Mousson and Port-sur-Seille, by 10 P.M. Progress was slow on the eighteen-mile journey owing to the poor condition of the track. A half hour was lost because of a minor derailment that occurred at about 7 P.M. The train was finally halted in the Bois de Boule at 1:30 A.M. on the twenty-first and the engine sent to Rosieres for repairs. Later that afternoon the engine returned, and the journey to the Bois de Facq resumed at 7 P.M. Two hours later the train reached its destination, and the tanks were unloaded and concealed in the woods.

Higgins's men spent most of the twenty-second performing maintenance on the tanks while the officers conducted a reconnaissance of the routes they would follow that night to Lesmenils, Morville, and Port-sur-Seille. Several French NCOs arrived that afternoon to act as platoon guides. At 7 P.M. they led the tanks out with orders to move through the French wire, maneuver around in no-man's-land for ten minutes, and return. The mission was accomplished without incident before midnight.[7]

Viner, meanwhile, received instructions to reconnoiter the French 125th Division sector for a similar mission in the vicinity of Sivry the night of 23–24 September.[8] This task was also accomplished without incident, and Higgins and his force were reunited with the rest of their battalion near Boureuilles on 27 September, the second day of the Meuse-Argonne offensive.[9]

During Viner's absence the first of a number of tank units organized in the United States began arriving at the 302d Tank Center. Late in the afternoon of 20 September the officers and men of the 331st Tank Battalion detrained at Brennes. They were followed on 22 September by the 328th, 329th, and 330th Tank Battalions, the 376th and 377th Training and Replacement Companies, and the 316th Repair and Salvage Company. By the end of October the force undergoing training at the tank center would be augmented by the 326th, 327th, 332d, 335th, 336th, and 337th Tank Battalions; the 302d, 304th, and 306th Heavy Tank Battalions; the 307th and 318th Repair and Salvage Companies; the 380th and 381st Training and Replacement Companies; the 303d Tank Center Headquarters; and several casual companies and individual replacements. This brought the strength of Viner's command to 656 officers and more than 8,000 men by 1 November.[10]

The demand for individual officer and NCO replacements at the front would siphon off skilled instructors from the nucleus Viner had retained at the center, making it "practically impossible," given the center's remaining limited resources, to train more than one battalion at a time. As all of the battalions arrived with little or no training, it was possible to train and field only one, the 331st, prior to the Armistice.[11]

Patton got his first look at the new operations plan on 15 September, when he met with Rockenbach in the St. Mihiel sector. The plan called for the First Army to kick off the attack at 5:30 A.M. on 26 September after a three-hour artillery preparation. The sector was divided roughly equally into three corps zones. The III Corps sector was on the right, bounded to the east by the Meuse River and a north-south line to the west—just east of the high ground at Montfaucon. The V Corps had the center sector, sharing its eastern boundary with III Corps to the right, and its western boundary, which was a north-south line east of Exermont, Cheppy, and Neuvilly, with I Corps on the left. The I Corps sector included the Aire River and the Argonne Forest in the western third of the First Army zone.

Each of the corps was allocated three assault divisions, and First Army had an additional fourteen divisions in reserve. The I Corps's 35th Division on the left; the V Corps's 91st, 37th, and 79th Divisions in the center; and the III Corps's 4th Division on the right were charged with making the deep penetrations on either side of Montfaucon. The I Corps's 28th Division had the secondary mission of driving a wedge into the sector east of the Argonne Forest and clearing the bluffs overlooking the Aire River. French forces on the western side of the forest had a similar mission. The intent of these latter maneuvers was to isolate German forces within the Argonne. The 77th and 37th Divisions were to conduct limited attacks into the Argonne, the purpose of which was to hold the Germans there in position. The 33d and 80th Divisions in the III Corps sector were tasked with establishing blocking positions along the Meuse to protect against a German counterattack from the east.[12]

Here Rockenbach, after reviewing a reconnaissance report prepared by one of Lt. Col. Emile Wahl's officers on 10 September, determined that the combination of streams, woods, high ground, and shell craters in the first three miles of the sector would make the area virtually impassable for tanks. Because of this, he thought they could be counted on only for operations north of the Bois de Montfaucon–Baulny line. Beyond that

point he saw three distinct avenues of approach for tanks—two in the
V Corps sector and one in the I Corps zone east of the Aire River. Based
on this analysis, Rockenbach allocated Pullen's larger 3d Brigade to the
V Corps and Patton's 1st Brigade to the I Corps. Pullen's French 505th
Regiment and two groupes of the heavier Schneiders and St. Chaumonds
would operate in the 79th and 37th Division sectors north of the Bois de
Montfaucon. The French 17th Light Tank Battalion with forty-eight
Renaults was later added to the 505th for the operation and assigned to
operate in the 91st Division sector. Pullen was additionally tasked with
ensuring the units he attached to the 79th Division assisted, "wherever
practicable," the III Corps advance.[13]

This secondary mission points out one of the weaknesses of the overall
plan: command and control. The decision to establish the III and V Corps
boundary east of Montfaucon meant close coordination would be required
between the two corps and the 79th and 4th Divisions as they conducted
their pincer movement around the high ground. This would have been
difficult under any circumstances, but it proved to be especially so in this
case because the 79th had yet to experience combat. In fact, more than
half the 79th's enlisted men, mostly from Pennsylvania and Maryland, had
been in the Army just four months.[14]

Rockenbach informed Patton that his brigade would retain its present
composition and operate in the I Corps sector in support of the 35th Division. Armed with this information, Patton set off by automobile on 16 September to conduct a personal reconnaissance. Dressed in a French
uniform, he gave the area near Vauquois Hill north of Clermont a thorough going-over. Here Patton discovered, to his delight, that the trenches
were much narrower and the ground in much better condition than they
had been near St. Mihiel.[15]

While Patton worked out his concept of operation, Brett, Compton, and
Chanoine prepared their units for the move to the new sector. Both of
Chanoine's Schneider groupes and Brett's 344th Battalion entrained at
Bernecourt of the night of 19 September, and Brett's tanks arrived in Cler-
mont the next night. While they were detraining, a German artillery bar-
rage hit the station, but, fortunately, the only thing damaged was nerves.
The battalion moved to an assembly area in a clump of woods about a
mile north of Clermont and west of the Clermont-Bourevilles road. The
area was under the direct observation of Germans located in the Vauquois
Heights, so on the night of Saturday the twenty-first, Brett, acting on a

First Army Plan of Attack, 26 September 1918

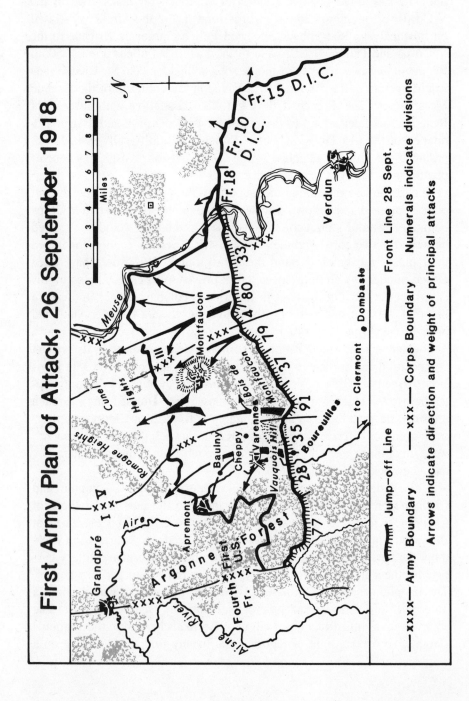

Legend:
- ꟷꟷꟷꟷ Army Boundary
- xxx ꟷ Corps Boundary
- Arrows indicate direction and weight of principal attacks
- ꟷꟷꟷ Front Line 28 Sept.
- Jump-off Line
- Numerals indicate divisions

hunch, moved the battalion about a half mile northwest to the Haute Prise Farm. It proved to be a wise decision, as the Germans heavily shelled the abandoned position early the next morning.[16]

Chanoine's tanks arrived without incident on the night of the twenty-first and were quickly shuttled from the rail station in Clermont to a neighboring woods to the west. But where was the 345th Battalion? Patton was particularly frustrated with Compton, whom he had found searching the Clermont area "for a house instead of getting his Tanks." He later wrote that he thought Compton was "a fool but I a greater one to trust him."[17]

Patton's lack of confidence in Compton's abilities was reflected in the operation plan he drew up for the brigade on the twenty-first. The plan also reflected an appreciation for the limits imposed by the terrain and the optimism implied in the First Army scheme of maneuver. Patton's plan called for employment of his brigade in depth in a narrow, mile-and-a-half-wide corridor extending northward from Clermont between the eastern edge of the Argonne Forest and the Bois de Cheppy near Varennes and Boureuilles. The Aire River served as the boundary between the two assault divisions, the 28th on the left and the 35th on the right, so the natural inclination would have been to allocate a single battalion to support each division's advance. Unfortunately, while the river worked well as a natural boundary between the two divisions, its position in the corridor precluded this option. The entire corridor was large enough to support a single tank battalion, but there was only enough trafficable ground in the 35th Division sector to support two companies. The 28th's sector had maneuver space for only one company.

Recognizing that Brett was the more reliable of his two battalion commanders, and that the 344th's tanks were worn out after their extended action in the St. Mihiel offensive, Patton decided to have the 344th Battalion lead the brigade, with A and C Companies trailing the forward elements of the 35th Division, and B Company west of the Aire River supporting the 28th Division's advance. Compton was instructed to array his companies in similar fashion but a distance of nearly a mile behind Brett's battalion. B and C Companies of the 345th were assigned to the 35th Division sector, and A Company to the 28th's. Chanoine's two groupes of cumbersome, inferior Schneider tanks were assigned the brigade reserve mission and would trail Compton by a little more than a mile until the ground became easier to negotiate beyond the I Corps objective.[18] Patton intended that Brett should support the advance as far as the first day's objective, then "leap-frog" the 345th Battalion forward to support the advance from that point on.[19]

Brig. Gen. Malin Craig, I Corps chief of staff, approved Patton's plan that same day.[20]

Two trains carrying fifty-six of Compton's tanks finally arrived at 1 A.M. on Tuesday the twenty-fourth. They were greeted by a barrage of German artillery fire, which had managed to zero in on the Clermont railroad station. One of the rail cars was struck just after a tank rolled off, and, to add to the confusion, the station's ramp broke with six of the battalion's tanks still sitting on the train. Capt. William H. Williams, the B Company commander, quickly jumped into the driver's seat of one of the tanks and drove it straight off the end of the car without the aid of a ramp. The crewmen in the remaining five tanks emulated Williams's bone-jarring feat, and soon the battalion was scurrying for cover in the woods northeast of the station, a reverse slope position hidden from the prying eyes of German artillery observers.[21]

With the battalions settled into their assembly areas, leaders and reconnaissance officers began reconnoitering routes to the jump-off line and studying the terrain over which the initial stages of the battle would be fought. Several of the tankers ventured into no-man's-land, where they found evidence of earlier fighting—the remains of French and German soldiers killed during clashes in the sector in 1914. "They lay undisturbed as they fell nearly four years before," Captain Semmes later wrote. "The clothing they wore and the arms they had carried were recognizable, but they were uniformed skeletons."[22]

Rockenbach, wishing to avoid the embarrassment of running short of fuel, as the Tank Corps had done at St. Mihiel, ordered two fuel dumps established near the front lines. One was set up at Dombasle to support Pullen's 3d Brigade. It contained 325,000 liters of gasoline, 32,000 liters of "fine" oil, 13,000 liters of Valvoline, 6,000 kilos of grease, and 9,000 liters of ordinary oil. Patton's petroleum supplies—30,000 gallons of gasoline, 1,000 gallons of heavy motor oil, 1,000 gallons of light motor oil, and 5,000 pounds of grease—were positioned at Recicourt.[23]

Patton went a step further. Still smarting from the vision of Brett at Nonsard with tears of frustration streaking his sooty cheeks and pleading for fuel so his battalion could continue the advance, Patton ordered that two twenty-liter gasoline cans be tied to the tail of every tank. While this "created some danger from fire [a bullet striking one of the gas cans could start a fire that would quickly torch the entire tank], . . . the risk was thought preferable to the lack of gas." Patton further supplemented the forward fuel supply by establishing fuel dumps less than a half mile from the front line at the Côtes de Forimont and the Abancourt Farm, located west of Boureuilles. He managed to stockpile an additional 20,000 gallons of gasoline at those two sites.[24]

Second Lt. George B. Heilner, the 344th's reconnaissance officer, was assigned the task of shepherding two truckloads of gas and oil to the Côtes de Forimont on the night of 23–24 September. As he and his detail were unloading the trucks, the Germans, who "happened to be particularly active" that night, laid down a barrage that gave everyone in the area "a most uncomfortable 15 minutes." Although caught in the middle of the artillery fire, Heilner and his men never missed a stride, completing the mission without further incident.[25]

Patton refined and published his field order on 24 September. The order included detailed guidance on routes and methods of advance, assigned graphic control measures, and gave specific instructions for supply and liaison operations. The brigade command post was established at the 35th Division's advance headquarters in the Côtes de Forimont. But Patton, despite Rockenbach's earlier displeasure with his failure to stay put at St. Mihiel, had no intention of remaining there. Patton figured he could keep his commander informed by taking a group of from six to ten runners with him as he advanced up Route Nationale No. 46 with the leading tanks. Patton likewise put the onus on battalion commanders to keep him informed. It was his hope that if they used "every means to get information [to higher headquarters, at least] some may arrive."[26]

Salvage and repair operations would be handled by the 321st Repair and Salvage Company based at Camp Fourgous, about a third of a mile south of Vraincourth.[27] Forward maintenance support was to be provided by a tank from each company specifically designated to carry extra fan belts and other minor repair equipment. The idea for this came from one of the brigade's privates, who observed during the St. Mihiel offensive that tanks suffering from minor malfunctions such as broken fan belts were spending too much time out of action because they had to be evacuated all the way to the rear for repair. Semmes later wrote that this idea served as the genesis for the tank company maintenance teams that were employed with such success in World War II and Korea and are still a part of the Army's tank battalion organization. That the idea came from a private in Patton's brigade should not be considered unusual. Semmes recalled that Patton was always asking junior officers and men for suggestions to improve the unit's abilities "because he knew that those intimately working and fighting with tanks would furnish worthwhile ideas."[28]

Patton made one last inspection of his battalions on Wednesday the twenty-fifth. He found conditions in Compton's battalion "very dirty" and gave the young commander a tongue-lashing before moving on to the 344th Battalion, where he found things "better but could stand improvement."[29]

That night the tanks were moved up to their departure positions—the companies east of the Aire into the trenches near Boureuilles-Vauquois, and those west of the river in the wooded area four kilometers north of the Abancourt Farm.[30]

On the eve of battle Patton could count a total of twenty-eight Schneider tanks in Chanoine's 14th and 17th Groupes, sixty-nine Renaults in Brett's 344th Battalion, and fifty-eight in the 345th. The 345th was still awaiting the arrival of Lieutenant Higgins and the fourteen tanks he had taken on the deception mission in the St. Mihiel sector.[31]

The first intimation the Germans had that something was brewing in the Meuse-Argonne sector came at 11:30 on the night of 25 September, as the American long-range artillery batteries began firing sporadic barrages into their rear areas. This harassing fire gave way to a thunderous, pounding, three-hour preparation at 2:30 A.M. on the twenty-sixth as the more than 2,700 artillery pieces gathered in the sector opened up. More than 100,000 doughboys crammed the front-line trenches, listening apprehensively to the almost constant whistling of rounds passing overhead or peering over the parapets to watch the brilliant spectacle as thousands of shells crashed into the German positions across the twenty-mile front.

At 5:30 A.M., H hour, the sound of whistles split the air, and sergeants, platoon leaders, and company commanders sprang from the trenches, calling for their men to follow them "over the top." Visibility was extremely poor. A dense fog shrouded the entire First Army zone, and artillery smoke added to the eerie effect as soldiers drifted in and out of the mist.

As expected, German resistance was light. Sporadic machine-gun fire and intermittent artillery bursts had little effect on the advancing Americans. Their real enemies were the fog, the shell-pocked landscape, the tangled undergrowth, and the barbed wire—all of which made it difficult to maintain contact with the protective artillery barrages rolling inexorably toward the German rear.

Brett's 344th Battalion moved out with the lead elements of the 28th and 35th Divisions in the I Corps sector. Because of the poor visibility, most of the company commanders and platoon leaders preceded their units on foot. Capt. Newell P. Weed, the Company B commander, and one of his runners walked steadily forward about 300 yards in front of the tanks and infantry, searching for places where the tanks could more easily cross the trenches. After losing contact with the tanks in the smoke and fog, Weed ordered his runner to set off in one direction to find them while he moved in another. A burst of machine-gun fire sent Weed diving into a nearby trench for shelter—and into the waiting arms of several German infantrymen. They quickly surrounded and disarmed him, then set off for the rear with their prize. The small group had gone only a short distance when one of Weed's tanks arrived on the scene. The Germans ordered Weed to be silent and not move, but he ignored their threats and was able to signal the tank, which advanced in their direction. Faced with this turn of events, "the Boche dropped their booty and took to their heels," and Weed was soon reunited with his company.[32]

While Weed was with the Germans, two of his men, 2d Lt. Edward Bowes and Cpl. Walter H. Blanchard, left their tank and crawled nearly 150 yards through a withering blast of machine-gun fire to rescue a wounded doughboy lying in front of trenches occupied by men of the 28th Division west of the Aire.[33]

The protective concealment offered by the smoke and fog allowed Weed's company, trailed by Capt. Dean M. Gilfillan's Company A, 345th Battalion in reserve, to quickly reach the vicinity of Boureuilles, about two miles forward of the jump-off line in the 28th Division sector. Unfortunately, German machine-gunners occupying Le Chéne Tondu, a rocky outcropping on the edge of the Argonne Forest overlooking the Aire River valley and the high ground near Vauquois on the opposite bank, set up a deadly cross fire with their comrades in the vicinity of Cheppy. As the German resistance stiffened, Gilfillan was forced to commit his tanks in support of Weed. Although the tanks initially reached Varennes at 9:30 A.M., the doughboys moving behind them were unable to advance through the deadly hail of machine-gun bullets, which became more and more accurate as the fog dissipated. The tanks were forced to fall back and join them.[34]

Gilfillan's tank was hit twice by artillery fire during the initial advance on Varennes, and it began to burn. Wounded by machine-gun fire, Gilfillan remained in his tank, destroying two German machine-guns and wreaking havoc on a column of enemy soldiers trying to reach the town. As the fire worsened and the threat of an explosion grew imminent, Gilfillan was

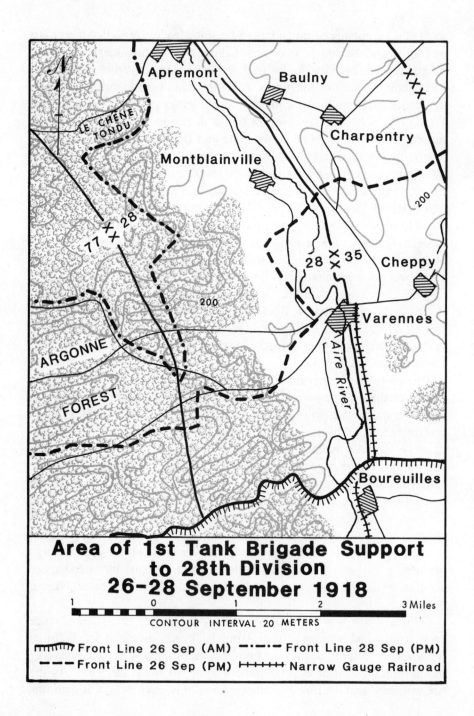

N

Apremont

Baulny

XXX

LE CHÊNE TONDU

Charpentry

Montblainville

200

77 XX 28

Cheppy

28 XX 35

200

Varennes

ARGONNE

Aire River

FOREST

Boureuilles

Area of 1st Tank Brigade Support to 28th Division 26-28 September 1918

1   0   1   2   3 Miles

CONTOUR INTERVAL 20 METERS

ꟼꟼꟼꟼ Front Line 26 Sep (AM)   ‒ ‧ ‒ ‧ ‒ Front Line 28 Sep (PM)

‒ ‒ ‒ Front Line 26 Sep (PM)   ┼┼┼┼┼ Narrow Gauge Railroad

forced to abandon the vehicle, only to be wounded a second time by fragments from a nearby shell burst. He stayed on the scene long enough to turn over command of the company to 1st Lt. Thomas G. Brown and to watch 2d Lt. Harry M. Mayne lead his platoon of tanks ahead of supporting infantry in a successful assault on the town.[35]

During the second assault on Varennes, the tank commanded by Sgt. Raymond C. Chisholm of Company A, 345th Tank Battalion was hit in the turret by artillery fire. Chisholm, fatally wounded by the blast, ordered his driver to keep moving forward while Chisholm continued to fire at German positions until he dropped dead at his post.[36]

The tanks advancing in the 35th Division sector moved with similar ease while the fog lasted. Capt. Math L. English's Company C, 344th Battalion was able to pass quickly through the much-vaunted Vauquois Heights positions overlooking the jump-off line. This area had suffered a tremendous pounding during the artillery preparation, leaving the Germans stunned and of little threat to the advancing infantry. The only real obstacle encountered during the early going was a mine field. But signs reading "Achtung! Minen!" left behind by the Germans made it possible for English's tankers to negotiate it without loss.[37]

Captain Semmes, who was trying to move Company A, 344th Battalion, through the low ground east of the Vauquois Heights, found the going more difficult. Second Lt. Edwin A. McClure, one of his platoon leaders, recalled that the area was a virtually impassable bog, riddled with shell holes and trenches and swept by artillery and machine-gun fire. Undaunted, Semmes set out on foot to reconnoiter a passage for his tanks. Despite suffering painful wounds as he moved forward through the intense German fire, Semmes was able to guide his tanks through the first line of trenches, but he soon discovered the terrain was impassable. Wounded a second time in the head and unable to continue, he placed 1st Lt. Leslie H. Buckley in command and instructed him to take the company back and head forward west of the high ground on the Clermont-Boureuilles-Varennes road.[38]

Second Lt. Guy Chamberlain, a platoon leader in Company C, 345th Tank Battalion, became the first Tank Corps officer to die in action when he was shot while walking ahead of his tanks, trying to lead them through a particularly difficult piece of terrain in the Vauquois Heights.[39]

Patton, who had told his subordinates he would remain in the brigade command post for at least an hour after the attack started, soon became

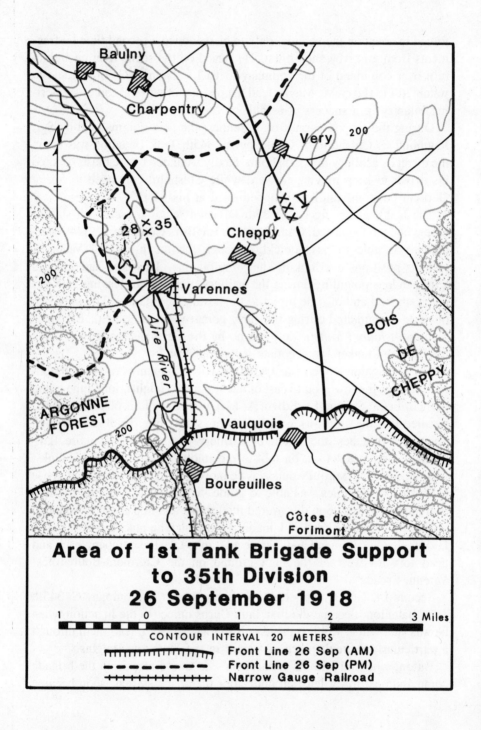

Baulny

Charpentry

Very   200

I XX V

28 XX 35

Cheppy

Varennes

Aire River

BOIS

DE

CHEPPY

ARGONNE
FOREST   200

Vauquois

Boureuilles

Côtes de
Forimont

Area of 1st Tank Brigade Support
to 35th Division
26 September 1918

1   0   1   2   3 Miles

CONTOUR INTERVAL 20 METERS

Front Line 26 Sep (AM)
Front Line 26 Sep (PM)
Narrow Gauge Railroad

impatient. He could hear the sounds of tanks moving, the crash of artillery fire, and the ragged sound of scattered machine-gun bursts. But he could see nothing through the thick mist. Unable to wait any longer, he set off between 6 and 6:30 A.M. with Captain Knowles; 1st Lt. Paul S. Edwards, the brigade signal officer; and about a dozen enlisted runners — some carrying field telephones and reels of wire and others baskets of pigeons. Patton, remembering the dressing down he had received from Rockenbach after St. Mihiel, fully intended to keep the Tank Corps's chief informed of his brigade's progress.

Patton and his party headed northward on the Clermont-Boureuilles-Varennes road leading toward Cheppy in the 35th Division sector. As they advanced, they passed by Chanoine's Schneiders and the Renaults from Compton's 345th Battalion operating east of the Aire River between Boureuilles and Varennes. The party finally halted near a crossroads just south of Cheppy, and Patton released one of the pigeons with a message reporting his location. [40]

It was here that the 35th Division's attack had faltered. The 35th, one of several untested divisions in the First Army line that day, had one of the toughest sectors along the twenty-mile front. The division's first-day mission would have given pause to veteran soldiers; for the National Guardsmen from Kansas and Missouri entering battle for the first time it proved to be too much. Enthusiasm can carry a soldier only so far. Good leadership is needed to keep up the momentum of the attack, and too many of the 35th's officers and NCOs fell victim to the intense German machine-gun fire. By the time Patton arrived on the scene, confusion reigned.

While Patton paused at the crossroads, several tanks from Company B, 345th Battalion arrived. They stopped briefly, then continued forward after the lieutenant in charge explained that they were moving up to try to relieve the pressure on the confused infantrymen, who were drawing fire from the front and flanks as well as from bypassed German positions to the rear. Alert German artillery observers spotted the tanks, however, and soon shells began landing nearby.

It was probably these tanks that came to the aid of members of the 138th Infantry Regiment's headquarters and staff, pinned down by German machine-gunners a short distance away. When the tanks arrived on the scene, the infantrymen, believing them to be French, asked the regimental interpreter to explain the situation to the crews. The interpreter, a Sergeant Morel from the French army, sprinted in front of one of the tanks, waving his arms and yelling. When the tank stopped, "the little trap door opened. Instead of the expected poilu [French soldier], an American soldier stuck his head out and asked calmly: 'Well, what the hell do YOU want?'"[41]

The German gunners had ceased fire in order to avoid giving away their positions, so Morel showed the tank commander where the enemy guns were hidden. Unfortunately, though the tankers tried hard, they were unable to get across an intervening creek. From the positions they were in, the tankers' machine guns proved unequal to the task of clearing the hill of the enemy. As artillery fire began to fall in the area, the 138th's commander, Col. Harry Howland, fearing the effects of the barrage would be greater than any further assistance the tanks could render, ordered them to pull back.[42]

Meanwhile, German machine-gunners had also spotted Patton's group, and he ordered everyone to take cover in a neighboring narrow-gauge railroad cut. Guards posted at either end of the defile soon reported that several groups of disorganized doughboys were walking or running toward the rear. Patton halted them and, after questioning, ordered them to join his group in the shelter of the railway cut. It soon became crowded, and Patton, concerned by the increasing tempo of the German fire, led his force, now numbering nearly 100, to the reverse slope of a small hill about a hundred yards to the rear. Here he ordered everyone to disperse and lie flat on the ground. No sooner was this accomplished than German machine-gunners began sweeping the area with fire—seemingly from all directions.[43]

Meanwhile, the two lead tanks from English's company had become stuck in a trench, barring the way to further progress for the rest of the unit. While English led his men in an attempt to dig the two tanks out, both of Compton's supporting companies and Chanoine's two groupes of Schneiders arrived, presenting a lucrative target for German artillery. Before the guns had a chance to register, however, the officers were able to disperse them to covered positions.[44]

Patton, seeing the activity, ordered Knowles to find Compton and tell him to get his tanks across the trench and to knock out the machine guns. When Knowles reached the trench, he discovered that two of the French Schneiders had bulled forward and become hopelessly stuck in the only other visible crossing point. He relayed his message to Compton and Captain Williams, the Company B, 345th Battalion commander, and returned to Patton's location.[45]

Growing increasingly angry at the seeming lack of response to his orders, Patton first dispatched Lieutenant Edwards, then Pfc Joseph T. Angelo, his orderly, to relay similar instructions to Compton. Patton would have gone himself, but he could see the infantrymen he had collected were beginning to panic, and he was afraid they would flee if he left them.

Edwards, meanwhile, had seen the difficulty the men were having in the trench and relayed his message to Captain English, whom he found standing near a platoon of tanks. Patton, by now thoroughly incensed, finally arrived on the scene and quickly took charge, organizing a concerted effort to clear the trench. He ordered the French tankers to begin digging out their Schneiders, then stomped over to English's Renaults, which were being "splattered with machine-gun fire." He started stripping off the shovels and picks strapped to their hulls and put every available man to work. Soon several score doughboys, tankers, and engineers were busily digging out the sides of the trench, clearing a passage for the vehicles. All the while the Germans poured heavy machine-gun fire into the trench system, and a steady stream of shells pounded the area. Although some twenty men were wounded, Patton and English stood on the parapet throughout the operation, directing the men's activities. In response to repeated requests that he seek shelter from the fire, Patton shouted, "To Hell with them — they can't hit me[!]"[46]

As soon as the obstacle was cleared, English led five tanks up the hill in the direction of Cheppy. As the lead vehicle crested the ridge, Patton leaped to his feet and rushed forward, waving a large walking stick that looked like a cane, and shouting, "Let's go get them! Who's coming with me?" About a hundred men responded to his call, leaping up to follow. Several members of Patton's command group would have joined them, but they were unarmed or occupied with other duties. Sgt. William V. Curran was burdened with a field telephone and wire, Sgts. L. T. Garlow and Lorenzo F. Ward had to guard the pigeon baskets, and Sgt. Harry M. Stokes was busily engaged in a shell hole bandaging a wounded doughboy's leg.[47]

As Patton and his pieced-together force followed the tanks over the crest, they were met by a withering blast of flanking machine-gun fire from the vicinity of the Varennes-Cheppy road. Everyone hit the ground, and Patton, unable to rally the group, asked for volunteers to assault the machine guns with him. Only seven men responded, and as they advanced toward the German gunners with Patton and Private First Class Angelo, they fell, one by one. About forty yards from the guns Patton was struck in the upper leg, and Angelo wrestled him into a shallow shell hole in the middle of the open field just a few hundred yards short of Cheppy. Angelo cut away Patton's trouser leg and bandaged the wound – a dangerous proposition, as every movement they made drew a burst of fire from the nearby German machine-gunners. A platoon from English's company, trying to catch up with its commander, soon arrived, causing the Germans to hold their fire in an effort to conceal their positions from the tank crews. In the ensuing lull, Patton sent Angelo to tell the tankers where the machine guns were located. Armed with this information, the tanks moved off to engage them.[48]

A short time later, one of the NCOs from the command group, a Sergeant Schemnitz, came along searching for the colonel. Angelo saw him and called him over to the shell hole, where Patton told Schemnitz to go back with word that he was wounded, that Brett was in command, and that no one should come forward to carry him back because they would draw enemy fire. Schemnitz, agitated by the news, rushed back to the command group shouting out Patton's instructions, then set off in search of a stretcher. Garlow released a pigeon with a message for Rockenbach's headquarters.[49]

Compton, meanwhile, had managed to get his tanks through the trench works and dispatched 2d Lts. Darwin T. Phillips and Clarence W. Cleworth around the west side of the hill with their platoons. Second Lt. Edward J. Mitchell led his platoon over the crest while their company commander, Captain Williams, acting on Compton's orders, followed behind. Capt. Courtney H. Barnard led Company C, 345th Battalion in a sweep around the east side of the hill.

After cresting the hill, Compton joined Phillips's platoon, which was busily rooting the Germans out of the trenches to the west. Once they cleared this obstacle, they were joined by a major from the 138th Infantry Regiment leading about 100 men. Together the tanks and infantry flanked Cheppy by moving northwest of the town. The remaining tanks from the 345th Battalion led the remainder of the 138th Infantry into the town, and by about 1:30 P.M. it was secure.[50]

Members of the 138th Infantry Regiment confirmed the effectiveness of Compton's tanks in the assault on Cheppy. When the tanks arrived on the scene, they

maneuvered perfectly, they swung out of column and into line, crossed the open between the two roads and took position about ten paces apart. All effectives in that neighborhood formed in squads behind the tanks. While the infantry with rifle and machine-guns fired on suspected places in the brush and woods on the hillside, the tanks, with all their armament, pounded the [machine-gun] nests and pill boxes to pieces. The one-pounders [37mm cannon] fired high explosives into the defenses and the machine-guns cut down the evicted gunners. It was but a work of minutes for these wheeled forts, and the front was cleared of an enemy which had delayed the advance for hours.[51]

With the machine guns silenced, Sergeant Schemnitz, Sfc. Earnest E. Ely, and Cpl. John G. Heming came forward with a stretcher. Together with Angelo, they loaded Patton onto the litter and carried him to the rear of the hill, where he reiterated his instructions to Lieutenant Edwards, who set off in search of Major Brett. Sgt. Edgar W. Fansler relieved Angelo on the stretcher, and the five soldiers carried Patton about a mile and a half to the rear, where they delivered him into the care of an ambulance company. Angelo remained with Patton while the others returned to their duties at the front. Patton, ever mindful of his duty, refused to be evacuated until he had personally reported to the 35th Division headquarters. An ambulance drove him there with Angelo at his side, and an officer from the division staff came out to talk with Patton. His report delivered, he allowed himself to be taken to a field hospital.[52]

Back in Cheppy, Compton, instructed by Patton to take charge of the tanks at the front until Brett could be located, consolidated his force south of town at about 3 P.M. Lieutenant Buckley arrived at about 3:30 with the nine tanks remaining in Company A, 344th Battalion, and Captain English showed up with only two tanks still operable. Major Chanoine's two Schneider groupes were reorganizing at the base of the hill that had been the scene of their earlier troubles.[53]

Compton set up his command post in a shell hole near the forward command post of the 35th Division. Brett still had not shown up, so Compton reported to Maj. Gen. Peter E. Traub, the 35th's commander, and explained that he was presently commanding Patton's tank brigade. After listening to Compton's report, Traub ordered him to send one company, preferably Chanoine's remaining Schneiders, toward Very to assist the infantry and the other French tanks that had managed to press forward with them. Chanoine's groupes were still in a state of confusion, so Compton sent Lieutenant Buckley forward with eleven Renaults—the remnants of the 344th Battalion's Companies A and C. This freed Captain English to go in search of the rest of his scattered command and attempt to reorganize it. Buckley left at once and was able to link up with the infantry in time to assault Very. Company A remained there overnight, joined later in the evening by Captain English, who came forward with the rest of Company C.[54]

Brett, who had stayed in his command post until 9 A.M. before setting out after his tanks, had his own difficulties getting forward. He spent most of the day with his party of three officers and ten enlisted runners dodging machine-gun and artillery fire and trying to catch up with his lead elements. Lieutenant Edwards finally found him in midafternoon, and he joined up with Compton in the early evening.[55]

The two companies supporting the 28th Division's attack on the west side of the Aire were far from finished for the day when they took Varennes. They made several attempts to drive the Germans out of their positions in the high ground northwest of the town. During one such attempt Cpl. Donald M. Call, the driver of a platoon leader's tank in Company B, 344th Battalion, barely managed to escape from his burning tank after half of its turret was blown off by a German 77mm shell fired from point-blank range. Call sought shelter in a nearby shell hole, but upon discovering that the tank commander, 2d Lt. John Castles, had failed to follow him, returned to the tank. Ignoring the intense artillery and machine-gun fire, Call dragged Castles, who was seriously wounded, out of the vehicle and carried him more than a mile to safety.[56]

First Sgt. Lester W. Atwood, the senior enlisted man in Company A, 345th Battalion, spent the morning serving as a runner. While seeking cover in a trench outside of Varennes, he encountered a tank driver and, finding out that his tank commander had been shot through the jaw and evacuated, saw a chance to get into the fight. The tank had been abandoned nearby, so Atwood set out with the driver, recovered the vehicle, which was undamaged, and joined up with three others from his company who were advancing westward from Varennes toward the Argonne. Atwood and his group skirted the forest, encountering little but occasional bursts of machine-gun fire. At one point, on a hill due west of the town, they came upon a pair of recently abandoned German 77mm guns, and Atwood ordered a brief halt. The tankers took advantage of the respite to top off their gas tanks with fuel from the twenty-liter cans strapped to the tanks and to check oil and water levels. They then set off in the direction of Apremont, the Corps's first-day objective.

Because it was getting dark, Atwood turned down the request of a 28th

Division company commander who wanted him to take his tanks and destroy some machine guns that had been harassing his troops. Instead, he halted the tanks, ordered them camouflaged, and set off on foot toward Varennes with Sgt. Sam Reese and a Sergeant Jones. Finding no one from the 1st Tank Brigade in town, they returned to their tanks after burying Sergeant Chisolm, who had been killed during the second assault on Varennes that morning.[57]

The French tanks attached to Lieutenant Colonel Pullen's 3d Brigade were far less involved in the fighting in the V Corps sector than Patton's brigade had been in the I Corps zone. There were two primary reasons for this. First, the massive artillery preparation so stunned the German defenders, already stretched thin in the forward trenches, that the assault divisions in the V Corps sector were able to make rapid progress. Second, the Bois de Montfaucon, a large wooded area, extended all the way across the Corps front. This required engineers to clear paths for the tanks—a laborious, time-consuming process that significantly retarded the progress of Pullen's tanks on the first day of the offensive.[58]

The advance to the Bois de Cuisy, a small woods southeast of the town of Montfaucon, was not entirely without incident, however. First Lts. Carl J. Sonstelie and Herbert J. Ellis, members of Pullen's staff, materially assisted the advance of the French tankers by moving forward through the woods on foot, seeking paths that could be cleared by the engineers. Both Ellis and Sonstelie repeatedly exposed themselves to German machine-gun and sniper fire while directing the engineers' efforts. At one point, Ellis routed a sniper who had been harassing the engineer detail he was supervising. Later, moving into the Bois de Cuisy with a French officer, Ellis helped kill seven Germans who were hindering the advance of the tanks.[59]

Like Patton, Pullen also found himself in the thick of action. During the attack on the Bois de Cuisy, Pullen rallied a group of disorganized doughboys from the 79th Division and led them through "violent machine-gun fire" to occupy the ground overrun by the tanks.[60]

All four battalions of French Renaults operating in the V Corps sector had cleared the Bois de Montfaucon by the evening of the twenty-sixth and had consolidated in the Bois de Cuisy. The heavier St. Chaumonds of the XI Groupement did not join them until the next morning.[61]

Patton's brigade lost a total of forty-three tanks during the first day's operations. Although more tanks had fallen prey to German fire than had in the St. Mihiel offensive, the tanks' greatest enemies were still mechanical failure, mud, and the deep German trenches.

As the sky began to lighten on the morning of 27 September, it became apparent that the hoped-for gains would not be attained. The Germans' stalwart First Guard Division had broken the back of the 35th Division attack the day before, leaving a shattered, confused mass of units trying to sort out the carnage. Part of the reason for this confusion was the friction between the National Guard and regular officers that existed in the division. Although this was a problem to some extent in many units, most were able to work around it. The problem was exacerbated in the 35th Division when the commanders of both infantry brigades, three of the infantry and one of the artillery regiments, and the chief of staff were relieved just a few days before the attack was to begin. Their replacements did not have sufficient time to learn their new jobs or to develop the bonds of trust and cooperation so vital during combat.

How bad was the confusion in the 35th's sector? Take as an example the experience of the commander of the 139th Infantry Regiment on the night of the twenty-seventh. He became lost while going forward to check the displacement of his lead elements. For the next twenty-four hours he wandered the battlefield in search of his unit, finally wandering into the command post of the 28th Division. While he was missing, his replacement was unable to find either the regimental command post or any of its subordinate units, the locations of which were unknown to either brigade or division. Meanwhile, the division's two brigades had become so badly intermingled that it was impossible to provide infantry units for support or maneuver.[62]

Rugged terrain rather than confusion was the problem on the west side of the Aire River valley. Looking northwest past Varennes along the twisting banks of the Aire, all one could see was a series of ravines stretching down from the high ground on the edge of the Argonne Forest, reaching like rocky fingers for the river. The Germans had set up a vicious defense in depth throughout this sector, emplacing machine guns in concrete pillboxes at the ends of the ravines and deploying artillery batteries in the high ground overlooking the valley.

Rockenbach, concerned over Patton's loss, ordered Brett to remain in the brigade command post and leave Compton in charge at the front. Brett in turn issued written instructions to Compton outlining the 1st Brigade's new command structure. In order to clarify the lines of authority in the forward units, Compton placed Captain Barnard in command of all tanks operating east of the Aire. This gave Barnard a force consisting of the remnants of Companies A and C of the 344th Battalion and Companies

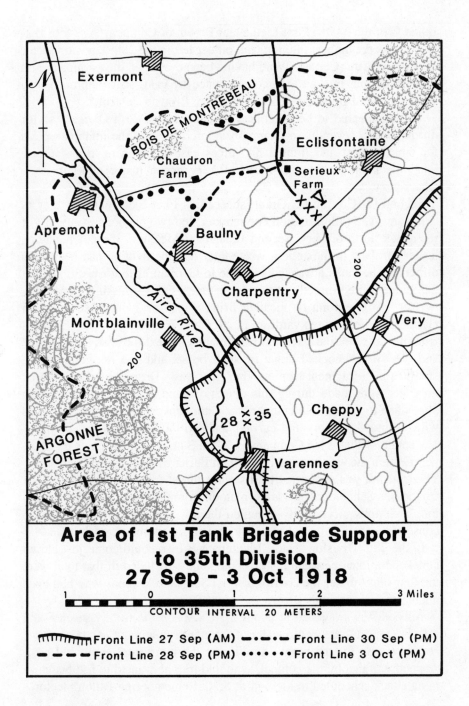

Area of 1st Tank Brigade Support
to 35th Division
27 Sep - 3 Oct 1918

CONTOUR INTERVAL 20 METERS

1   0   1   2   3 Miles

| Front Line 27 Sep (AM) | Front Line 30 Sep (PM) |
| Front Line 28 Sep (PM) | Front Line 3 Oct (PM) |

B and C of the 345th. First Lieutenant Brown of Company A, 345th Battalion, was the highest ranking tank officer left in the 28th Division sector west of the Aire (Captain Weed having been evacuated the evening before after stumbling into an area contaminated by gas), so Compton placed what remained of Weed's company under Brown's control.[63]

There appeared to be no concerted effort in the 35th Division sector on the twenty-seventh, so Compton ignored most of the infantry's pleas for tank support and limited his forces east of the Aire to assisting a local attack on the plateau north of Very and a major effort to seize Charpentry.[64]

Second Lt. Gordon M. Grant, succeeding First Lieutenant Buckley as Company A, 344th Battalion commander, led two platoons in support of the attack north of Very. Second Lieutenant McClure, the other platoon leader involved in the action, wrote later that the fighting had been very difficult. The tanks advanced so close to German artillery positions that the gunners fired directly at them over open sights—missing them "only by feet." At one point he recalled firing a load of 37mm case shot at a German soldier just ten yards away. McClure frequently had to dismount from his tank "to find out what [the infantry's] next move was involving," and was twice knocked down by shell bursts and had his helmet brim smashed by a fragment from another near miss. Despite repeated efforts, no gains were made during this operation, and after waiting until long after dark for a possible counterattack, the tankers loaded their vehicles with wounded soldiers and carried them to the aid station at Very.[65]

Captain Williams led Company B, 345th Battalion, in an attack on Charpentry from the west, trailed by Captain Barnard's Company C in reserve. Their assault was supported by a similar thrust from the south by Major Chanoine's Schneider tanks and a platoon from Company C, 344th Battalion. The tanks successfully occupied the town, but the infantry failed to consolidate their gains, choosing to dig in on the outskirts instead.[66]

In the 28th Division sector, Lieutenant Brown could count just eleven tanks operational between his two companies. Included in that total were the four tanks under the control of First Sergeant Atwood, who had been unable to link up with Brown the night before.

Atwood was awakened at about 2:30 A.M. on the twenty-seventh and asked to report to the commander of the infantry regiment in whose sector he had set up for the night. Arriving at the regimental command post, Atwood was met by the colonel, who told him he wanted to use his tanks in an attack scheduled to kick off at 5:30. Remembering Patton's lectures

at Bourg, Atwood decided his first responsibility was to help the
infantry—orders or no orders. At 5 A.M. he moved his small element into
positions designated by the regimental commander and, as the 28th Division's artillery had not been able to move forward during the night,
provided the only artillery preparation available—a "barrage of cast iron
one pounders" from his two tanks equipped with 37mm guns.

The attack was postponed until 6 A.M., when it became light enough
to see about fifty yards ahead. At that time Atwood led his four tanks in
a line formation spread out at intervals of about a hundred meters and
moving at a rate of about fifty meters per minute. The Germans responded
with a withering blast of machine-gun, sniper, and antitank fire. Within
minutes the other three tanks had fallen behind, leaving Atwood alone at
the head of the assault, "routing out several Boche and chasing them way
over to the woods." His tank began to lose power, so Atwood returned to
friendly lines, where he discovered that an antitank bullet had penetrated
the tank's armor and holed the radiator, causing the engine to overheat.
While his driver remained with the damaged tank, Atwood set off to find
out what had happened to the other three tanks. He learned that two of
the tanks had experienced minor mechanical problems that the crews had
since corrected, and that Sergeant Reese had been killed in the third tank,
although it was still operational. Atwood commandeered Reese's tank and,
together with the other two, skirted the edge of the Argonne Forest, routing more Germans "out of their lairs." By the end of the day the attack
had succeeded in reaching Montblainville, where Atwood was reunited
with Lieutenant Brown and the rest of his company.[67]

Brown's experience had been similar. As he and his seven tanks pressed
forward, they encountered intense machine-gun fire from concrete pillboxes. It was quickly discovered that the best way to handle them was to
fire high-explosive 37mm shells at the portholes, and in this way Brown
and his men were able to silence the German gunners.[68]

Sergeant Charles C. Young, one of Brown's tank commanders,
epitomized the fighting spirit of Patton's tankers. Twice wounded by fire
from antitank guns as his platoon led 28th Division doughboys through
a hail of machine-gun and artillery fire on an assault north of Montblainville near Apremont, he refused orders to be evacuated to the rear and
continued to fight his tank.[69]

Despite the arrival of Lieutenant Higgins with his fourteen tanks from
the St. Mihiel sector on the night of the twenty-seventh, the 1st Tank Brigade could count only eighty-three operational vehicles on the morning

of 28 September. Little attempt was made to maintain unit integrity at this point. Divisions would submit requests for a specific number of tanks to support their operations, and Brett and Compton would do the best they could to comply.

On 28 September the 28th Division asked for ten tanks; the 35th wanted forty; and the 91st Division, the unit sharing the corps boundary with the 35th Division, requested five tanks to serve in a liaison capacity. Lieutenant Brown was able to muster fifteen tanks from his two companies to support operations west of the Aire, Captain Barnard headed a force of forty-two tanks (including six French Schneiders) in the 35th Division sector, and Lieutenant Higgins was dispatched with a tank platoon to perform the liaison mission for the 91st. The remaining tanks were retained in brigade reserve.[70]

As German reinforcements began arriving in the First Army sector, resistance stiffened, a fact reflected in the 1st Brigade's losses on the twenty-eighth. Fifty of the tanks that went into action were lost—many to enemy fire, even more to mechanical failure or ditching.

All five of Lieutenant Higgins's tanks were put out of action within the first hour of combat in the 91st Division zone. Both Higgins and his driver were blinded when a German machine-gunner fired a well-aimed burst at their vision slits at point-blank range. Higgins's blindness was only temporary, however, so he changed positions with the driver and drove the tank toward the rear. Unable to see clearly, he drove into a trench, ditching the tank and forcing the pair to continue on foot until they were picked up and taken to a first-aid station.[71]

Barnard's element fought a pitched battle in the vicinity of the Bois de Montrebeau. Confusion still reigned in the 35th Division sector, and Lieutenant McClure, who was later awarded the Distinguished Service Cross for his unsupported action near Jonville during the St. Mihiel offensive, again found himself operating well in advance of the infantry. Undeterred, he dismounted to lead his platoon into the thickly wooded terrain, continuing the advance in the face of intense German fire. During the approach to the woods his platoon came across two Renaults that had been wrecked the previous day, and they stopped long enough to assist with the evacuation of the dead and wounded crewmen as well as several wounded infantrymen they found nearby.[72]

Not far away, Cpl. Charles V. Williams, a tank commander in Company B, 345th Battalion, left his damaged vehicle during an assault on a hedgerow south of the woods and charged through a barrage of artillery, rifle, and machine-gun fire to administer first aid to his wounded platoon leader. He then took the officer's place and led the platoon on to the objective.[73]

The battle for Montrebeau Woods was particularly bloody. The Germans, in response to the threat posed by the tanks, had supplied front-line units with antitank rifles and reinforced the forward positions with light artillery pieces. The guns were employed in a direct fire role against the tanks, often firing at point-blank range over open sights. It was into this maelstrom that the tankers, supporting an attack by doughboys of the 140th Infantry Regiment, plunged.[74]

Second Lt. Julian K. Morrison, a platoon leader in Company A, 344th Battalion, was forced to dismount and lead his five-tank element more than 400 meters on foot through intense enemy fire that knocked three of his vehicles out of action. After consolidating in the woods, he continued to press the attack with his remaining two tanks, leading a force of 35th Division doughboys as far as Charpentry before darkness halted them.[75]

Farther east, 2d Lt. John W. Gleason, one of Barnard's platoon leaders, led his five tanks in advance of a battalion of infantry. The platoon knocked out numerous dug-in German machine-gun positions, clearing the way for the doughboys, who were able to secure the Bois de Montrebeau.[76] Barnard's force was able to press the attack as far as the high ground to the east and west of the woods, taking the towns of Baulny and Chaudron, and the Serieux farm. He was forced to pull back, however, when the infantry withdrew to a more defensible position.[77]

Lieutenant Brown, leading the fifteen tanks operating in the 28th Division sector, was ordered to seize the town of Apremont. Murderous machine-gun and artillery fire from the high ground along the eastern edge of the Argonne Forest again halted the supporting infantry, and Brown was forced to pull back. His tanks assaulted Apremont five times during the day before the infantry was finally able to occupy the town. Patton wrote later that it "seemed [the infantry had] forgotten the fire power which they themselves possessed and expected the tanks to completely obliterate all resistance before they would advance."[78]

On the morning of 29 September Major Chanoine was ordered to withdraw his mechanically inferior Schneider tanks from the I Corps sector. The French tankers had fought well, but their machines had reached the limit of their endurance and would require extensive maintenance and repairs before they would be able to return to the front. This left Brett and Compton with a force of fifty-five Renaults. The figure would have been significantly less without the efforts of the battalion maintenance crews and the officers and men of the 321st Repair and Salvage Company, who worked throughout the night to recover ditched tanks and correct mechanical problems.[79]

Both the 28th and 35th Divisions attempted to continue the attack that day, but the doughboys, exhausted and hungry after three continuous days of battle on short rations, had about reached the limit of their endurance. Forty tanks were again committed to support the 35th Division, and Captain English led the remnants of Companies A and C, 344th Battalion, in an attack out of Baulny. The Germans were believed to have employed gas in the sector, so English's tankers moved out with their protective masks on. They were able to press the attack as far as Exermont, northwest of the Bois de Montrebeau but had to pull back when, once again, the infantry withdrew.

First Lt. Harvey L. Harris, who had been called forward as a replacement from the 331st Tank Battalion, which was undergoing training at Bourg, wrote later of this action:

About 200 meters from [the] juncture of [the] National Highway, I could see that the Boche were shelling the junction with G.I. cans [slang for heavy artillery]. A Captain, who I knew was on the staff of the 35th Division, stopped me and said, "Thank God, we can't advance on account of machineguns." It wouldn't have made any difference to him [if the] opposition was 420's. He'd have wanted our little tanks to clean them up anyway. It's surprising what they ask us to do. Doughboys to Generals have sent us against places a battleship couldn't capture. . . . "There are snipers along the road. You'd better get in your tank," he cautioned. I walked a little further and saw a couple of doughboys go down — then decided to get aboard. It isn't any easy matter for me to get into the tank through those two small rear doors even when in training — let alone with a haversack, pistol, belt, etc. I was hung up for a while but got in and lumbered down the road. There didn't seem to be anything going on so I climbed out and sat just back of the turret, my feet hanging inside and my body almost entirely protected in front by the turret. I passed the General and his staff sitting in a gully at the side of the road. I'll never forget the appreciative "You're the man we're waiting for" expression as I rode by.[80]

Harris led his tanks as far as the road junction and was "jarred several times from [the] concussions of exploding shells" during the advance. The company commander appeared and motioned for Harris to deploy the tanks behind a nearby hedge. After the tanks were camouflaged in position, the crews moved to the shelter of a nearby gully and watched an attack in progress in the 28th Division sector across the river:

> There I was safe from every shell and, looking out across the river [saw tanks] advancing in line, infantry following at several hundred meters—just as pretty as the shows we'd put on at Bourg for the staff school. At the far edge of the flats was the east edge of the Argonne. Everything seemed to be going well until a shell exploded amidst the infantry. Then another in front of the advancing tanks; then another to the right. The doughboys immediately flopped down and the tanks continued, not knowing it. Then I saw a puff of smoke where forest sort of jutted out towards me and a shell exploded within ten feet of a tank. Several turned for a depression near [the] river, the others went for the woods, not knowing where the shots were coming from. And there I sat unable to do a thing, probably the only one who knew where those guns were, the flash had told me. With the aid of glasses I could actually see the Boche gun crews working, loading, and if my voice would have carried, I could have yelled the instant the #1 (man who fires the piece) had his hand on the grip and was going to fire.[81]

Harris was distraught over the fact that there was nothing he could do. Moments later a field artillery captain approached his vantage point accompanied by two enlisted men. They were there to set up a forward observation post for the officer's battery.

"See anything to shoot at?" the captain inquired.

"Well, I'll be damned," Harris replied, pointing at the German gun positions across the river.

Within minutes the officer was able to call in a barrage on the enemy guns, scoring direct hits on at least two of the pieces and wreaking havoc among the gun crews.[82]

Confusion in the ranks of 35th Division units east of the woods caused them to fall back, giving up some of the ground taken at such heavy cost the previous day. Compton, who had held Barnard in reserve with the remaining elements from Companies B and C, 345th Battalion, ordered the tanks to advance north of Charpentry and northwest of Very, and the line was restored.[83]

English and his tanks were withdrawn to the Bois de Montrebeau to help resist a counterattack expected in the center of the division's sector. English, supported by some of the division's engineer troops, was ordered to hold a line extending from Baulny to Eclisfontaine. The attack failed to materialize, and the tankers and engineers held their ground while the infantry formed a new line of resistance to their rear.[84]

In the 28th Division sector, Lieutenant Brown divided his fifteen tanks into two groups led by 1st Lt. Harry E. Gibbs and 2d Lt. John W. Roy. Their elements attempted to work northward from Apremont but were forced to withdraw in the face of intense artillery fire. As in the 35th Division sector, an expected counterattack never developed, and the line remained the same.[85]

All along the American front units collapsed in exhaustion. Pershing's ability to control the First Army had deteriorated, and efforts to supply the front-line forces had reached a critical state. Only three roads served the entire twenty-mile front, and men, machines, and animals became perpetually snarled in massive traffic jams. Casualties being evacuated to the rear frequently had to wait more than a day before they could be moved back as trucks and horse-drawn wagons carrying much-needed food and ammunition for the units at the front inched forward, picking their way around crews filling holes and otherwise trying to shore up the fragile roadways.[86]

The 1st Tank Brigade was withdrawn to reserve positions in the vicinity of Montblainville in the 28th Division sector and Baulny and Charpentry in the 35th Division zone on the morning of 30 September for badly needed maintenance and reorganization. At 6 P.M., orders were issued for the relief of the badly mauled 35th by the veteran 1st Infantry Division. The Kansas and Missouri National Guardsmen had suffered between 6,000 and 8,000 casualties during the frightful four-mile advance up the Aire River valley.

Late in the evening of the 30th, the 28th Division asked for eight tanks to support a local attack early the next morning. Two platoons under the command of 2d Lt. Harry D. Heitz, Company B, 345th Battalion, were dispatched to Apremont and spent the night preparing for the 6 A.M. assault. The Germans beat them to the punch, however, striking a counterblow at 5:30 on the morning of 1 October. Fortunately, the tanks were in position and ready to fight. During the ensuing battle five tanks were destroyed by artillery fire, and Sgt. Earling F. Dutt, Cpl. James M. Whitney, and Pvt. James Casey, all of Company A, 345th Battalion, were killed. Several others were seriously wounded. Lieutenant Heitz's tank was one of those destroyed, but both he and his driver escaped without serious injury.[87]

The attack was repulsed, and prisoners taken by the infantry later told 28th Division intelligence officers that the tanks had "treated them like ironing clothes. They passed through their lines, turned and ran over them. They never had any experience with tanks."[88]

Repair and salvage operations continued around the clock despite the proximity of the German lines and the intense artillery and machine-gun fire. Second Lt. Don C. Wilson, a maintenance officer in the 321st Repair and Salvage Company, joined the many victims of the German artillerymen in the heights overlooking the Aire. While he was making a parts run to the maintenance crews working near Charpentry, his motorcycle was bracketed by shell bursts on the road outside of Varennes. Recalling the incident later, he wrote:

> The hun had direct observation over about three kilometers of that road ... [and] I am sure that he was shooting at me, for I was the only one on [it] and they [the shells] followed me, it may have been that he was strafing the entire road at the same time, but I think not. I fooled with the throttle a few moments until without my knowing it I had turned it open. I never was a speed maniac, but that was one time in my life when I appreciated it.
>
> I had barely passed Fritz's line of sight and was slowing down gradually when one of his gas shells burst in the road just ahead of me. I couldn't stop so I had to go through it, but held my breath for some time afterwards. ... It only made me cough slightly, but two days later I was evacuated with the Flu.[89]

Wilson's hospital stay was brief, however. It took him only two days to tire of the hospital routine, and after hearing that his company commander, Capt. Ellis Baldwin, had also been evacuated, Wilson went AWOL from the hospital and returned to his unit.

Thanks to the efforts of the officers and men in the forward maintenance teams, the 1st Brigade could count eighty-nine tanks operational by the night of 3 October.

Patton, although wounded early on the first day of the offensive, had proven himself to be an outstanding combat leader. More importantly, he had instilled his own indomitable sense of duty and aggressive spirit in the men he led. Despite his absence from the battlefield, the officers and men of the 1st Tank Brigade continued to press on as though he had been there. The unit was a model of organization and proficiency, quickly adapting to the fluid changes on the battlefield, battering away at a formidable opponent, accepting heavy losses, and coming back again and again for more. This first generation of American tank crewmen truly earned the nickname "Treat-'em-Rough Boys" with which they had been tagged.

This is the true measure of leadership: to be able to teach the skills needed to perform the required missions and to imbue one's subordinates with the motivation to carry on in the absence of superiors and in the face of overwhelming odds. There is no doubt that Patton had that quality—nor was it lacking in the officers and men he trained. When Patton went down, Brett assumed the mantle of command. Lieutenants quickly took over for wounded captains, sergeants and corporals for wounded and dead lieutenants. Privates pressed on despite the absence of any constituted authority.

On the eve of the renewal of the offensive on 4 October, Rockenbach submitted a report to Pershing's headquarters outlining the status of Tank Corps units in support of the First Army. Casualties (both dead and wounded) in the 1st Tank Brigade amounted to 53 percent of the officers, including the brigade commander, four of the six captains, twelve of the twenty-four lieutenants, and sixty-five enlisted men (25 percent).

Based on the number of tanks available and assuming a daily advance of five kilometers, Rockenbach estimated that he could keep seventy-two tanks operational for action on 5 and 6 October, forty-eight by the seventh, thirty-two on the eighth, sixteen on the ninth, and the brigade would be "finished" by the tenth. He added that Chanoine's Schneider groupement was "mechanically broken down and returned to depot."[90]

Rockenbach projected a force of two St. Chaumond groupes would be ready to support the III Corps attack on the fourth and fifth, that a third groupe would become available on the sixth, but that attrition would totally deplete this force by the eighth. In the V Corps sector Pullen would be able to field two French light tank battalions on D day, these to be joined by two more by the evening of 6 October. Rockenbach figured only one battalion would be left in fighting condition for operations on the eighth and ninth, and, as with the 1st Brigade, the 3d would be combat ineffective by the tenth.[91]

It was hardly an optimistic report, and the officers and men of the Tank Corps resolved to prove him wrong.

Notes

1. Coffman, *War to End All Wars,* 303; Millett and Maslowski, *For the Common Defense,* 355; Maurice Matloff, ed., *American Military History* (Washington, D.C.: Center of Military History, U.S. Army, Publication 30-1, 1985), 399.
2. Coffman, *War to End All Wars,* 300.
3. Ibid., 301.
4. Ibid.
5. Patton, "304th Brigade History," 25.
6. Ibid., 26. The 326th and 327th Tank Battalions began using their new designations, 344th and 345th, respectively, following the St. Mihiel offensive.
7. Ibid., 26–27.
8. Commanding general, Second Army (U.S.), to commanding general, 125th Division (French), 21 Sept. 1918, Viner Collection, Document no. 27, World War I Survey Collection, Archives, USAMHI.
9. Patton "304th Brigade History," 27–28.
10. Lt. Col. Joseph W. Viner, "History of Tank School from August 20th, 1918 until Its Close," 14 Jan. 1919, Modern Military Records Division, Record Group 120, Entry 22, Folder 229, NA.
11. Ibid.
12. Coffman, *War to End All Wars,* 304–5.
13. Rockenbach, "Tank Corps Operations," 18.
14. Coffman, *War to End All Wars,* 305.
15. Rockenbach, "Tank Corps Operations," 19; Blumenson, *Patton Papers,* 601.
16. Patton, "304th Brigade History," 28; Weed, "Personal Experience Report."
17. George S. Patton, Jr., to Beatrice Ayer Patton, 22 Sept. 1918, as quoted in Blumenson, *Patton Papers,* 606.
18. Patton, "304th Brigade History," 29–30.
19. Rockenbach, "Tank Corps Operations," 20; Col. George S. Patton, Jr., "Operations of the 304th Brigade, Tank Corps from September 26th to October 15th 1918," 18 Nov. 1918, Patton Chronological Files, Box 11, 7–20 Nov. 1918, Patton Collection, 3.
20. Patton, "304th Brigade History," 29.
21. Ibid., 28–29; Compton, "345th Battalion War Diary."
22. Semmes, *Portrait of Patton,* 51.
23. Rockenbach, "Tank Corps Operations," 19–20.
24. Patton, "304th Brigade History," 31.
25. Heilner, "Personal Experience Report."
26. As quoted in Blumenson, *Patton Papers,* 608.

27. Rockenbach, "Tank Corps Operations," 20; Patton, "Operations of the 304th Brigade, 26 September–15 October 1918," 4.

28. Semmes, *Portrait of Patton,* 52.

29. George S. Patton, Jr., diary entry, 25 Sept. 1918, as quoted in Blumenson, *Patton Papers,* 608.

30. Patton, "304th Brigade History," 31.

31. Patton, "Operations of the 304th Brigade, 26 September–15 October 1918," 1–2. Pullen's French 505th Regiment and the attached French 17th Battalion, both equipped with Renaults, and the attached groupes of St. Chaumonds and Schneiders contributed an additional 250 tanks to the First Army zone ("Appendix 7, Tank Corps Operations"). This means that 419 tanks of various types supported the First Army in the Meuse-Argonne – a considerably larger force than the 189 Renaults erroneously cited by other historians who have previously dealt with this campaign. The discrepancy was most likely the result of failure to count all of the French-manned vehicles when determining the number of tanks supporting the First Army.

32. Weed, "Personal Experience Report"; General Order no. 24, GHQ – Tank Corps, AEF.

33. General Order no. 24, ibid.; General Order no. 2, General Headquarters – Tank Corps, AEF, Subject: Award of Distinguished Service Cross, 11 Feb. 1919, Patton Chronological Files, Box 11, 11–28 Feb. 1919, Patton Collection.

34. Patton, "304th Brigade History," 33; Weed, "Personal Experience Report"; Compton, "345th Battalion War Diary"; Laurence Stallings, *The Doughboys: The Story of the AEF, 1917–1918* (New York: Harper & Row Publishers, 1963), 243.

35. Compton, "345th Battalion War Diary"; General Order no. 24, GHQ – Tank Corps, AEF.

36. General Order no. 24, ibid.

37. Blumenson, *Patton Papers,* 610; 2d Lt. Arthur Snyder, "Company 'C' 344th Battalion, 2d Platoon, Argonne Sector," undated, Patton Military Writings, Box 47, Personal Experience Reports of Tank Operations – 1918, Patton Collection.

38. General Order no. 24, GHQ – Tank Corps, AEF; McClure, "Personal Experience Report"; Compton, "345th Battalion War Diary."

39. Brett, "Report of Major Brett."

40. Capt. Maurice H. Knowles, "Report of Circumstances of Wounds Received by Col. G. S. Patton, Jr.," undated, Patton Chronological Files, Box 11, 17 Dec. 1918.

41. Clair Kenamore, *From Vauquois Hill to Exermont: A History of the Thirty-Fifth Division of the United States Army* (St. Louis, Mo.: Guard Publishing Co., 1919), 117–18.

42. Ibid., 118.

43. Knowles, "Report of Circumstances"; 1st Lt. Paul S. Edwards, "Gallant and Exemplary Conduct of Col. Geo. S. Patton, Jr., and the Circumstances Leading Up to His Being Wounded in the Argonne Attack—September 26, 1918," 27 Nov. 1918, Patton Chronological Files, Box 11, 17 Dec. 1918, Patton Collection; Patton, "304th Brigade History," 32.
44. Patton, "304th Brigade History," 32.
45. Knowles, "Report of Circumstances."
46. Edwards, "Gallant and Exemplary Conduct"; Knowles, "Report of Circumstances"; Patton, "304th Brigade History"; Blumenson, *Patton Papers*, 611–12.
47. Edwards, "Gallant and Exemplary Conduct"; Knowles, "Report of Circumstances"; Blumenson, *Patton Papers*, 612–13.
48. Patton, "304th Brigade History," 32–33; Blumenson, *Patton Papers*, 614.
49. Edwards, "Gallant and Exemplary Conduct"; Blumenson, *Patton Papers*, 614–15.
50. Compton, "345th Battalion War Diary."
51. Kenamore, *Vauquois Hill to Exermont*, 119.
52. Blumenson, *Patton Papers*, 615.
53. Compton, "345th Battalion War Diary."
54. Ibid.
55. Heilner, "Personal Experience Report"; Edwards, "Gallant and Exemplary Conduct."
56. General Order no. 2, GHQ—Tank Corps, AEF, Subject: Award of the Medal of Honor; Semmes, *Portrait of Patton*, 54.
57. Atwood, "Personal Experience Report."
58. Rockenbach, "Tank Corps Operations," 29.
59. General Order no. 24, GHQ—Tank Corps, AEF.
60. Ibid.
61. Rockenbach, "Tank Corps Operations," 29.
62. Coffman, *War to End All Wars*, 311–12.
63. Compton, "345th Battalion War Diary"; Weed, "Personal Experience Report."
64. Compton, "345th Battalion War Diary"; Patton, "304th Brigade History," 33–34.
65. McClure, "Personal Experience Report."
66. Patton, "304th Brigade History," 34; Compton, "345th Battalion War Diary."
67. Atwood, "Personal Experience Report."
68. Patton, "304th Brigade History," 34; Compton, "345th Battalion War Diary."
69. General Order no. 24, GHQ—Tank Corps, AEF.
70. Rockenbach, "Tank Corps Operations," 22; Patton, "304th Brigade History," 35; Compton, "345th Battalion War Diary."
71. Patton, "304th Brigade History," 35.

72. McClure, "Personal Experience Report"; General Order no. 2, GHQ–Tank Corps, AEF, Subject: Award of Distinguished Service Cross.
73. General Order no. 2, ibid.
74. Kenamore, *Vauquois Hill to Exermont,* 181–82.
75. General Order no. 24, GHQ–Tank Corps, AEF.
76. Ibid.
77. Compton, "345th Battalion War Diary."
78. Patton, "304th Brigade History," 35–36; Rockenbach, "Tank Corps Operations," 22; Compton, "345th Battalion War Diary."
79. Patton, "304th Brigade History," 36.
80. 1st Lt. Harvey L. Harris, "Personal Experience Report," 19 Dec. 1918, Patton Military Papers, Box 47, Personal Experience Reports of Tank Operations– 1918, Patton Collection.
81. Ibid.
82. Ibid.
83. McClure, "Personal Experience Report"; Patton, "304th Brigade History," 37; Compton, "345th Battalion War Diary."
84. Patton, "304th Brigade History," 37.
85. Ibid.; Compton, "345th Battalion War Diary."
86. Coffman, *War to End All Wars,* 313–14.
87. Compton, "345th Battalion War Diary."
88. Ibid.; Patton, "304th Brigade History," 38.
89. 1st Lt. Don C. Wilson, "Personal Experience Report."
90. Rockenbach, "Tank Corps Operations," 23.
91. Ibid., 23–24.

A maintenance team works to free a 326th/344th Battalion Renault from a trench where it ditched near Nonsard on 13 September 1918. Leather helmets were standard for AEF tank crews.

Doughboys of the 35th Division ride 1st/304th Tank Brigade Renaults into action near Boureuilles on 26 September 1918.

A damaged 1st/304th Tank Brigade Renault awaits recovery near Varennes during the Meuse-Argonne campaign.

A 326th/344th Tank Battalion Renault passes through Exermont on its way to the front. The soldier in the foreground looks at the remains of a dead German soldier mutilated by the passing tanks.

Renaults from the 326th/344th Tank Battalion advance with soldiers from the 35th Division near Boureuilles on the first day of the Meuse-Argonne campaign, 26 September 1918.

French tank crews prepare to off-load their Renaults from a train on 25 September 1918. They supported the American 80th Division in the V Corps sector the following day.

Doughboys examine the wreckage of a 1st/304th Tank Brigade Renault and supply wagon. The vehicles were damaged by a German mine near Fléville on 12 October 1918.

German mock-ups of a British heavy tank and a Renault light tank. The Germans used cloth, wood, and scrap metal to construct these dummy tanks, which were used in the Meuse-Argonne campaign to deceive Allied aviators.

Damaged 1st/304th Tank Brigade Renaults await recovery near the German machine-gun position they destroyed during the Meuse-Argonne campaign.

A 1st/304th Tank Brigade Renault crosses a bridge repaired by 28th Division engineers near Boureuilles on 28 September 1918.

1st/304th Tank Brigade vehicle commander receives instructions before moving out in support of 28th Division troops near Varennes on 26 September 1918.

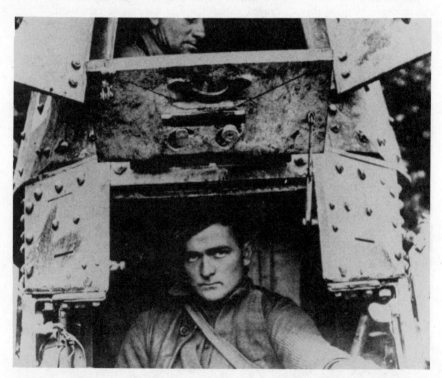

Sgt. Edward White and his driver, Cpl. Edward J. Elliott, 326th/344th Tank Battalion, open their hatches and get a breath of fresh air during a lull in the Meuse-Argonne drive.

A soldier examines the wreckage of a 301st Heavy Tank Battalion Mark V destroyed by a mine near Ronssoy on 29 September 1918.

A 301st Battalion Mark V roars through the shattered streets of Bellicourt on the way to rejoining its unit at the front on 3 October 1918.

A column of 301st Battalion Mark Vs rolls toward the front to support the 27th Division in the Selle River attack on 17 October 1918.

Damaged 301st Battalion heavy tanks await repair near Ronssoy after participating in the 29 September 1918 assault on the Hindenburg Line in the St. Quentin Canal region.

CHAPTER VIII

The Meuse-Argonne,
Phases Two and Three:
War of Attrition

As dawn approached on the morning of 4 October, the doughboys in the First Army's three corps waited anxiously in their trenches for the order to launch a general attack that, it was hoped, would crack the Hindenburg Line and allow them to seize the Army's 26 September objectives. Maj. Gen. Robert L. Bullard's III Corps had the mission of securing the Cunel Heights in the eastern third of the First Army sector. In the center, Maj. Gen. George H. Cameron's V Corps, whose battle-weary 37th, 79th, and 91st Divisions had been replaced by the veteran 3d and 32d Divisions, was assigned the task of capturing the high ground northwest of Romagne. Maj. Gen. Hunter Liggett's I Corps, still plagued by the Germans doggedly hanging on to their well-prepared positions in the Argonne bluffs overlooking the Aire River, was ordered to clear them out and secure the ridges north of Exermont. They would be led by the veteran 1st Infantry Division, which had replaced the badly mauled 35th Division three days earlier.[1]

Rockenbach's two tank brigades retained their original assignments, the 1st still in support of the I Corps on the left and the 3d's headquarters continuing to act as liaison for the French tank units supporting III and V Corps operations. Major Brett, with eighty-nine tanks operational in the 1st Tank Brigade, allocated two companies to support the 1st Division's assault and one to the 28th Division west of the Aire River. The remaining

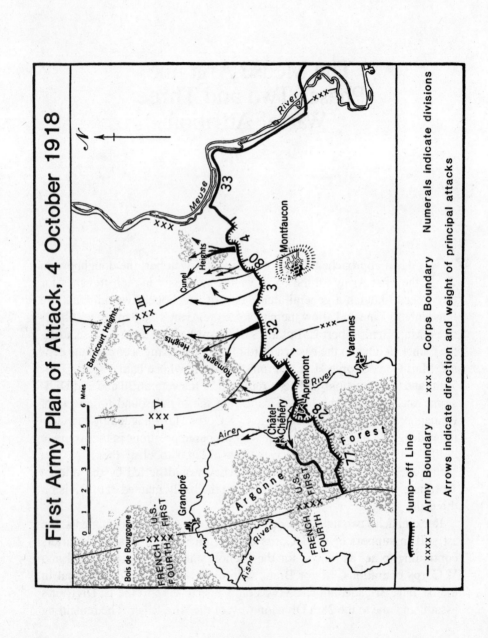

First Army Plan of Attack, 4 October 1918

N

Meuse River

Montfaucon

Cunel

Heights

Barricourt Heights

Romagne Heights

Varennes

Apremont

Aire River

Châtel-Chéhéry

Grandpré

Bois de Bourgogne

Argonne Forest

Aisne River

U.S. FIRST

FRENCH FOURTH

FRENCH FOURTH

U.S. FIRST

33

4

80

3

32

I

28

2

77

III

V

V

I

XXX

XXX

XXXX

XXX

XXX

XXX

XXX

0 1 2 3 4 5 6 Miles

xxxx — Jump-off Line

— xxxx — Army Boundary   xxx — Corps Boundary   Numerals indicate divisions

Arrows indicate direction and weight of principal attacks

tanks were held in brigade reserve.[2] Lieutenant Colonel Pullen's 3d Tank Brigade was able to muster only nine St. Chaumonds from the French XI Groupement to support the III Corps attack. Two French Renault-equipped battalions were committed in the V Corps zone — the 15th in support of the 3d Division and the 17th with the 32d.[3]

At 5:25 A.M. the American line again surged forward through the swirling autumn mists headlong into withering bursts of machine-gun fire and violent barrages of artillery. Sgt. Arthur Snyder, a tank commander in 2d Lt. H. A. Wood's platoon of Company C, 344th Battalion, recalled that only he and Lieutenant Wood were able to reach the 1st Division's jump-off line south of Exermont, the other tanks from their element having gotten lost en route to the point of departure. They were immediately taken under fire, and Lieutenant Wood was wounded in the face by shell fragments that penetrated his tank's vision slits. As Wood ordered his driver to head for the rear, a 77mm shell struck Snyder's tank, breaking one of its treads. He and his driver abandoned the vehicle and were able to crawl back to Wood's tank, which Snyder commandeered after sending the officer back to the rear for treatment. Snyder continued forward, but was soon forced to turn back when his driver suffered wounds similar to Wood's. He found a replacement driver, a Private Hadley from Company A, 344th Battalion, and quickly returned to action. Snyder and Hadley passed the lead infantry elements and pressed their attack well into the German line before both were wounded and the tank stalled after a German antitank bullet pierced its radiator. The pair remained in the vehicle, where they were rescued later in the day by elements of the 18th Infantry Regiment attacking between Exermont and Fléville.[4]

The tanks operating near the Bois de Montrebeau under the command of Capt. Math L. English of Company C, 344th Battalion, were slowed by especially difficult terrain. English dismounted from his vehicle and made a personal reconnaissance of the area in the face of heavy machine-gun and artillery fire. Moving on foot, he led his tanks forward until he was killed by a burst from a German machine gun.[5] Following English's death, 1st Lt. Robert C. Llewellyn continued to press the attack in this sector, leading his platoon in an assault against a battery of German 77mm guns firing directly at them at close range. The attack was successful, but Llewellyn was killed instantly by a direct hit that destroyed his tank.[6]

Second Lt. Edwin A. McClure of Company A, 344th Battalion, kicked off the attack with a mad dash through a barrage of German artillery fire as he led his platoon in advance of elements of the 1st Division's 16th Infantry Regiment. Continuing the assault, McClure's platoon cleared out numerous German machine-gun positions while scaling the high ground northwest of Exermont. As the infantry advanced, enfilading fire from a machine-gun nest located across the division boundary in the 28th Division sector pinned them down. McClure seized the initiative and crossed the boundary to take out the position, a maneuver that cost him his tank when both he and his driver were wounded, the sights on his own weapon were shot out, and its protective shield, damaged by enemy fire, jammed the gun's elevation mechanism. McClure dismounted from his tank and, dodging machine-gun bursts, ran to another, trading places with the wounded sergeant commanding it. As McClure continued to lead his platoon forward, an antitank bullet penetrated the side of his tank and struck two 37mm cartridge cases, igniting the powder, which burned his face and several maps he had stuffed into the ammunition rack. He quickly put out the fire with his bare hands before it could explode the remaining rounds. McClure's driver stalled the vehicle, and while it was stopped it continued to take fire from an antitank rifle. McClure was able to locate the gunner and put him to flight with 37mm fire.

A German 55mm gun opened up as the platoon moved up another hill. McClure knocked it out, and the combined fire from his two remaining tanks drove a platoon-sized German element out of hiding and sent them running toward the rear. He attempted to pursue them but was held up by an impassable stream just short of the crest. While they were halted, several German 77mm guns opened fire at close range, destroying the other vehicle. McClure ordered his driver to move in a zigzag pattern to a covered position, where he was met by a liaison officer from the 16th Infantry who ordered him to wait while friendly artillery was called in on the German battery and the doughboys consolidated their position.

McClure later linked up with other members of the platoon who had managed to commandeer several undamaged tanks. He tried to lead them back to Exermont to join forces with other members of the battalion but had to give up the attempt when they encountered heavy German artillery fire that included numerous gas shells. He finally left the platoon sergeant in command and set off on foot to search for his company or battalion commander and to seek medical attention for his wounds. While walking down the road toward Baulny, McClure passed out and was picked up by a truck headed for Cheppy. There he "escaped" from a mobile hospital and was able to locate a Tank Corps liaison officer in the 1st Division headquarters. He was finally evacuated after reporting the location of his tanks.[7]

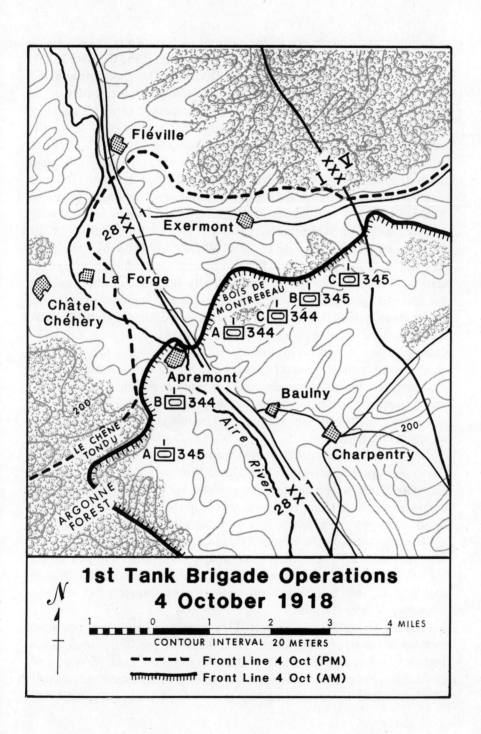

Fléville

XXV

28 XX I

Exermont

La Forge

BOIS DE
MONTREBEAU

C 345

B 345

C 344

A 344

Châtel
Chéhèry

Apremont

Baulny

B 344

200

Aire River

A 345

Charpentry

200

LE CHÊNE
TONDU

28 XX I

ARGONNE
FOREST

1st Tank Brigade Operations
4 October 1918

N

|—1——0——1——2——3——4 MILES|

CONTOUR INTERVAL 20 METERS

- - - - - Front Line 4 Oct (PM)

⊓⊓⊓⊓⊓⊓⊓⊓ Front Line 4 Oct (AM)

Fighting was also heavy in the vicinity of Charpentry, where Capt. Courtney H. Barnard led the remnants of Companies B and C, 345th Battalion, in support of elements of the 1st Division. The tanks encountered intense German fire as they moved forward from the northeast edge of the Bois de Montrebeau. Sgt. Harold J. Ash of Company C drove his tank forward through a barrage of 77mm artillery fire until it took a direct hit. Unable to advance, he and his tank commander, Sgt. Harley R. Nichols, remained with the vehicle. Nichols continued to fire on a nearby German machine-gun position until he was sure he had killed the crew. The pair dismounted from the tank and set out on foot toward the enemy position. As they moved forward, the Germans again opened fire on them. They were able to crawl close to the machine-gun nest and used their pistols to kill the two gunners occupying the position. While Nichols disabled the guns, Ash shot and killed two snipers. Finally, under the watchful eyes of another tank crew, they started back toward friendly lines. Along the way they encountered two Germans with antitank rifles and captured them.[8]

Not far away, Cpl. Albert J. Zimborski, also of Company C, ran his tank into a wooded area in an effort to rout entrenched German machine-gunners. The tank was hit by artillery fire while executing this maneuver and began to burn. Zimborski continued to fire on the Germans until the excessive heat forced him to abandon the vehicle. Both he and the driver were wounded while dismounting, but they decided to try to take the position on foot anyway. While charging the German gunners, Zimborski was hit again and killed.[9]

The 345th Battalion tankers continued the assault in the direction of Exermont. As they advanced, Sgt. David Winton, a tank commander in Company C, saw a German machine-gun nest in a nearby wood. Like Zimborski, he ran his tank into the trees, where it was hit and set on fire. Both he and the driver were shot while exiting the vehicle, then wounded again while trying to reach the Germans dismounted. The driver was struck a third time, as was Winton while attempting to assist him. The pair finally made it to a covered position, where they remained until nightfall. Winton then set out in search of help, receiving three more wounds before reaching the safety of friendly lines.[10]

First Lt. Thomas G. Brown was able to muster a force of sixteen tanks from the remnants of Company A, 345th Battalion, and Company B, 344th Battalion, to support operations in the 28th Division sector. Second Lts. Edward Bowes and John W. Roy led two platoons in advance of the infantry, which was able to take and occupy the towns of La Forge and Châtel-

Chéhéry. First Lt. Harry E. Gibbs led a platoon in support of an infantry assault on the German positions in Le Chéne Tondu, losing four of his five tanks to intense fire from the artillery batteries and "literally hundreds" of machine-gun positions dug in there.[11]

By the end of the day it became apparent that the First Army attack would not achieve its goals. Bullard's III Corps was stymied by intense fire from the German positions on the commanding high ground east of the Meuse River, and fierce German resistance slowed movement in the other two corps sectors. Only the 1st Division had made any appreciable gains, advancing about a mile and a half to the outskirts of Fléville, where it was held up by enfilading fire from the high ground west of the Aire River. Much of the credit for the division's success was attributed to the "effective work" of the 1st Brigade's tanks. Brett responded to this accolade by observing that the tankers' efforts would not have been nearly as good without the "splendid" support they received from the veteran doughboys.[12]

But the day's fighting had been very costly for the tanks operating in the I Corps sector. By the morning of 5 October only thirty could be counted ready for action. These were divided equally between the 1st and 28th Divisions, and only a single platoon, led by Lieutenant Gibbs, saw action that day. Gibbs, supported by elements of the 28th Division, led another attempt to root the Germans out of their positions in Le Chéne Tondu. During the attack, which was met by a fierce blast of artillery, mortar, and machine-gun fire, three of the five tanks were put out of action, and it faltered after gaining just fifty yards. Sgt. William E. Brophy, a tank commander in Company A, 345th Battalion, was wounded in the arm during the failed effort. He refused to return to the dressing station for treatment, and when word was passed around that the Germans were counterattacking, he insisted on returning to his tank, from which he put up a fierce fight until the attack was repulsed.[13]

Similar failures were experienced in the III and V Corps sectors, where the heights at Cunel and Romagne remained in German hands.

On 6 October the situation in the 1st Tank Brigade was becoming desperate. The tanks, which had been running constantly for nearly a month without an overhaul, were in terrible mechanical condition, with only seventeen considered ready for action. These were ordered held in reserve while the I Corps held its lines and supported an attack by the 82d Division's 164th Brigade in an effort to relieve the pressure on the 77th Division's "Lost Battalion" in the Argonne Forest.[14]

The "Lost Battalion" was not a single battalion but a force consisting of some 650 men from the 1st and 2d Battalions, 308th Infantry Regiment; the 3d Battalion, 307th Infantry Regiment; and the 306th Machine Gun Battalion. This small force had been cut off from the rest of the 77th Division since 2 October. Under the command of Lt. Col. Charles W. Whittlesey, commander of the 1st Battalion, 308th Infantry, and the ranking officer present, the doughboys had managed to hold out despite spirited German attempts to capture them. By the time they were relieved on 7 October, only 195 men remained unharmed by enemy fire.[15]

Although none of the 1st Tank Brigade's vehicles were involved in attacks on 6 October, one man died. Cpl. Harold W. Roberts of Company A, 344th Battalion, was moving his tank to a maintenance collection point near the Bois de Montrebeau when he drove into a "water tank trap" while trying to avoid enemy fire. As the vehicle overturned and began sinking, he told the gunner, "Well, only one of us can get out; out you go," and shoved him through the escape hatch. Roberts drowned.[16]

On 7 October none of the 1st Brigade tanks located east of the Aire River was in operable condition, and only eight could be mustered to support the 28th Division's attack west of the river. One of these struck a mine just minutes after going into action near Apremont, but the remainder continued to operate throughout the rest of the day—used mostly to support patrols as far as Châtel-Chéhéry.[17]

No tanks participated in operations from 8 to 10 October as maintenance crews worked around the clock trying to restore as many as possible to fighting condition. By the eleventh, forty-eight tanks were reported as being operable, and a force of twenty-three tanks under the command of Captain Barnard was dispatched from Varennes at midnight on the tenth to support an attack by the 164th Brigade of the 82d Division. The brigade had requested only five tanks, but because of the distance involved, Captain Compton decided to send as many as possible to ensure that the required platoon reached Fléville. Only four tanks managed to reach the link-up point with the 164th, and two more broke down shortly after the attack started, requiring the platoon to withdraw from the action, as Rockenbach had issued an order that tanks not be committed in less than platoon strength.[18]

This action marked the end of the 345th Battalion's participation in the campaign. Compton was relieved of his duties as commander of the brigade's forward elements on 12 October and ordered to assemble his men at Varennes in preparation for evacuation to the tank center at Bourg. Similar instructions were issued to the 344th Battalion. Both units left their tanks and detachments of men behind to be organized into the 1st Provisional Tank Company under the command of Captain Barnard. This force, officially created on the thirteenth, consisted of ten officers and 148 enlisted men and was equipped with twenty-four tanks, a Dodge automobile, four trucks, a motorcycle with sidecar, a rolling kitchen, and a water trailer. The 321st Repair and Salvage Company and Headquarters Company, 344th Battalion, remained at Varennes under the command of Major Brett to support the Provisional Company's operations.[19]

In Pullen's 3d Brigade, operations were marred by the lack of experience of the infantry divisions, which had never before operated with tanks. Problems were also encountered with liaison, and the French tanks were finally withdrawn from the sector on 10 October after a week of disappointing results.[20] During the campaign, the French tank units lost one officer killed, one officer missing, and sixteen wounded. Enlisted casualties amounted to nineteen dead, twenty-one missing, and 109 wounded. Twenty-five French tanks were lost to artillery fire and two to mines. Another twenty-two Renaults were abandoned by the French tankers behind German lines.[21]

By 10 October the Argonne Forest had been secured, but attacks by the III and V Corps had failed to make much headway. All across the American front, divisions were poised for a final assault on the Hindenburg Line. The plan Pershing and his staff devised for the operation called for a pincerlike maneuver by the V Corps's 5th and 42d Divisions aimed at isolating the Côte Dame Marie, a crescent-shaped ridge believed to be critical to the success of the operation. The veteran but weary 32d Division, which had been in the battle since the first of the month, was assigned the mission of assaulting the ridge three hours before the other divisions jumped off, in hopes of fixing the Germans in their positions so the two fresh divisions, which had been in the line only two days, could maneuver around their flanks.[22]

D day for the operation was set for 14 October, and Brett was ordered to meet with the new V Corps commander, Maj. Gen. Charles P. Summerall, who had succeeded Cameron on the twelfth. Summerall directed Brett to prepare the Provisional Company to support the 42d Division's effort between St. Georges and Landres-et-St. Georges at 6 the next morning. Brett sent word to Barnard, who had moved the Provisional Company from Varennes to Exermont, then went to the 42d Division headquarters to work out the details of the operation with Maj. Gen. Charles T. Menoher, the 42d's commander.[23]

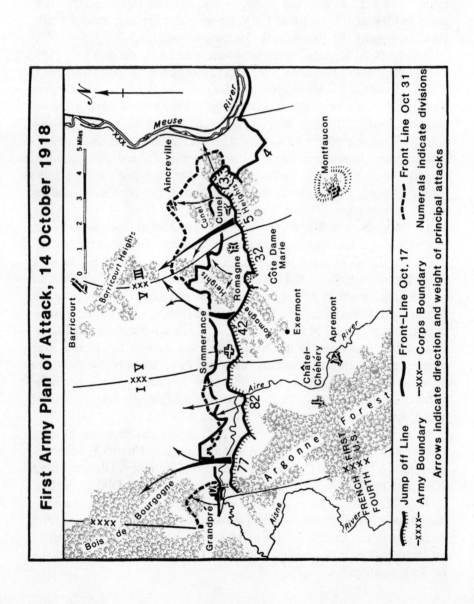

First Army Plan of Attack, 14 October 1918

Legend:

- ⌇⌇⌇ Jump off Line
- ━━━ Front-Line Oct.17
- ┅┅┅ Front Line Oct 31
- ╼╼╼ Army Boundary
- ×××× Army Boundary
- ××× Corps Boundary
- Numerals indicate divisions
- Arrows indicate direction and weight of principal attacks

The Provisional Company moved out for the front on the night of 13–14 October with its entire complement of twenty-four tanks, but because of "the long distance of the march, the road jams, and the grueling speed" the tanks were required to maintain to reach the point of departure, only ten made it in time for H hour.[24]

The tanks, led by 2d Lts. Lester W. Atwood (who had been commissioned on 5 October in recognition of his efforts while serving as first sergeant of Company A, 345th Battalion), Geddes, and C. C. Peters, advanced rapidly ahead of the infantry, who were slowed by the intense German machine-gun and artillery fire. Atwood later wrote that "'Jerry' seemed to have an especial grudge against the slits in the tanks for he played on [them] with his machine-guns in such a manner that the inside of the tank looked more like a Fourth of July celebration than anything else."[25]

The tankers penetrated the German line as far as Somerance, where they encountered what appeared to be a German force massing for a counterattack. They were able to bring sufficient firepower to bear to scatter the enemy "with considerable loss" before being forced to withdraw because of a lack of infantry support.[26]

The accurate German fire accounted for several casualties among the tank crews, including Atwood, who was hit in the arm and hand by splash; his driver, a Corporal May, who was shot in the mouth; and Geddes, a Corporal Williams, and a Sergeant Jones, all of whom were struck in the eyes by splinters from machine-gun fire aimed at the vision slits of their tanks.[27]

The overall success of the attack was disappointing. The flanking moves by the 5th and 42d Divisions were frustrated by the fierce German resistance, although both did make slight gains. The 32d Division, originally intended as "bait" to fix the Germans in position, was more successful, however, managing to secure the Côte Dame Marie by the night of 14 October. After two additional days of hard fighting, Brig. Gen. Douglas MacArthur's 83d Brigade of the 42d Division was able to capture the Côte de Chatillon, the other dominant ridge in the Romagne Heights, driving a significant wedge in the Hindenburg Line.[28]

The mid-October push was Pershing's final operation as a field army commander. Because of the unwieldy size of the First Army—it had grown to more than a million men and controlled some eighty-three miles of front—Pershing had earlier decided to create a second field army. On 12 October he appointed Lt. Gen. Robert L. Bullard to command the Second Army and gave him responsibility for a thirty-four-mile sector stretching

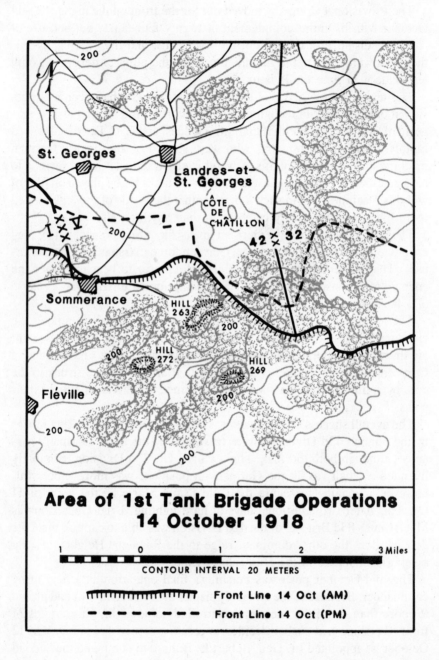

200

N

St. Georges

Landres-et-
St. Georges

A
CÔTE
DE
CHÂTILLON

I XX V

200

42 XX 32

X

Sommerance

HILL
263

200

200

HILL
272

200

HILL
269

Fléville

200

200

200

Area of 1st Tank Brigade Operations
14 October 1918

1   0   1   2   3 Miles

CONTOUR INTERVAL 20 METERS

Front Line 14 Oct (AM)

Front Line 14 Oct (PM)

from Fresnes to Port-sur-Seille in the east. Maj. Gen. John L. Hines, commander of the 4th Division, succeeded Bullard as the III Corps commander. Lt. Gen. Hunter Liggett took command of the First Army on 16 October after turning over the I Corps to Maj. Gen. Joseph T. Dickman on the fourteenth.[29]

The 1st Tank Brigade's Provisional Company was withdrawn to Exermont following the action of 14 October, where it remained until the end of the month, resting, performing maintenance, and helping scour the battlefield in search of disabled tanks that had been abandoned by their crews. All but one of the 141 tanks with which the brigade had begun operations on 26 September were located and recovered during this period.[30]

On 17 October orders were issued promoting Patton, Pullen, and Mitchell to colonel. Major Viner was promoted to lieutenant colonel.[31] Patton was still in the hospital recovering from his wound, and Pullen had returned to Bourg with his headquarters staff to begin organizing a new brigade consisting of the 331st and 345th Battalions, 306th Heavy Tank Battalion, and 316th Repair and Salvage Company.

Capt. John R. Johnson, an ordnance officer, was detailed to examine maintenance operations in the 1st Brigade on 2 October. He reported to the front on the fourteenth and spent a week inspecting the activities of the 321st Repair and Salvage Company at Varennes. Johnson found a great deal of confusion, as units had failed to record the location of abandoned vehicles, forcing maintenance personnel to comb the entire battlefield in search of them. He was especially critical of the original plan to have battalion maintenance sections perform all "running repairs." Although both Brett and Compton had heaped praise on their maintenance personnel, Johnson was sharply critical, claiming that the battalion organizations "had fallen down." He noted that in the "general confusion attending the advance [in] the St. Mihiel sector" much of the brigade's supplies, tools, and spare parts had been lost. Later, during the first phase of the Meuse-Argonne campaign, battalion commanders had ordered their maintenance sections to drag their remaining maintenance material with them into action, then forced them to abandon much of it when it held up the advance. When it finally became apparent that the forward maintenance sections could no longer cope with the vehicle losses, the 321st was ordered to perform all salvage and repair activities. Johnson praised the work of the officers and men of the 321st, adding that they had done "all that was humanely [sic] possible to keep the organization moving, under the circumstances and with what was left of the equipment."[32]

A running count maintained by Major Brett between 18 and 31 October shows the extent of the problem faced by the brigade's maintenance personnel. On the eighteenth, eighty-nine tanks were in the 321st's vehicle park at Varennes or had been located on the battlefield and were awaiting recovery. An additional thirty-one were forward with the Provisional Company at Exermont. One tank was in the process of being salvaged, and twenty tanks remained unaccounted for. Of the 121 tanks accounted for, fifty were in operable condition. On the nineteenth, nine more tanks were located by recovery teams; the total operable tanks remained at fifty. Three more tanks were found on the twentieth, and maintenance personnel had restored an additional three to fighting condition. By the twenty-third, 137 of the brigade's 141 Renaults had been located and the number in operating condition brought up to sixty-five. During the remainder of the month all but one of the brigade's tanks were recovered. Maintenance crews continued to work on those vehicles they had returned to Varennes and by the thirty-first had brought the number of tanks in operable condition to eighty-two.[33]

Patton, meanwhile, was discharged from the hospital on 28 October and returned to Bourg. What he found was not to his liking, and he issued a directive on the twenty-ninth ordering the men in the units returned from the front as well as those undergoing training at the center to improve their appearance standards and to strictly observe military courtesy at all times. He was particularly incensed at the manner in which many officers and men executed hand salutes:

> There is a widespread and regrettable habit in our service of ducking the head to meet the hand in rendering a salute. This will not be tolerated. In rendering salutes the head will be held erect. The hand will be moved smartly to and from the head-dress or forehead, if uncovered.[34]

Lieutenant Colonel Viner had done an admirable job with the limited training resources available to him at the 302d Tank Center. In just a little more than a month he was able to train the 331st Battalion, which joined the 345th Battalion and the 306th Heavy Tank Battalion as part of Colonel Pullen's 3d Brigade being formed at Bourg during the last week of the month. Although no additional tanks had been provided by the French or British, the 3d Brigade received orders on 29 October to move to Varennes and relieve the 1st Brigade. Only the two light tank battalions and the 316th Repair and Salvage Company would make the trip, as no heavy tanks were available. The order called for the 3d Brigade to take over the 1st's equipment when it reached Varennes.[35]

The Tank Corps was not alone in needing time to rest and reorganize after the three weeks of bitter fighting that culminated on 16 October. The entire First Army was in a state of exhaustion. Pershing, as commander in chief of the AEF and First Army commander, had been under constant pressure from Marshal Ferdinand Foch and Field Marshal Sir Douglas Haig to continue assaulting what they believed to be comparatively weak German defenses. One of Liggett's first acts as First Army commander was to ask for a two-week respite in order to relieve battle-weary, understrength divisions in the line and tidy up the front in preparation for a general attack in late October. His arguments convinced his superiors to grant him the needed time, and 28 October was set as the date for the renewed offensive.

Casualties had been extremely heavy. On 19 October Brig. Gen. Hugh A. Drum, the First Army chief of staff, estimated that more than 119,000 men had been killed or wounded in the Meuse-Argonne fighting. Although this was an overestimate, the depleted condition of the divisions at the front—several with less than 5,000 combat effectives—bore mute testimony to the severity of the action. Because replacements were not available, Pershing's headquarters ordered seven divisions broken up and their troops used to fill the First Army's thin ranks. In addition, Pershing directed that the strength of rifle companies be reduced from 250 to 175 men, forcing commanders in units whose casualties had been light to share their excess personnel with less fortunate comrades.[36]

Stragglers were also employed as replacements. Some men became separated from their units during the confusion of the fighting. Others reported being slightly gassed and hoped for evacuation or tried to get on details headed to the rear—from which they failed to return. Liggett believed the total number of stragglers in the First Army zone might be as many as 100,000. In order to tap this ready source of manpower, military police and special details scoured the rear areas, checking aid stations, field kitchens, and dugouts. As fast as the men could be rounded up they were returned to their units, some wearing humiliating placards with the inscription "Straggler from the Front Line" on their backs.[37] The 1st Tank Brigade, despite Patton's preachings about duty and its excellent record in combat, was not immune from having its share of shirkers. Second Lt. Charles H. Phillips, assigned to the 316th Repair and Salvage Company in the 3d Brigade, recalled overhearing shortly after his arrival at Varennes several lieutenants suffering from "battle fatigue" discussing the best place to get wounded so they could avoid further action.[38]

During the preparation for the renewed offensive, Liggett devised a plan similar to Pershing's 26 September concept of operation. The terrain, with heights in the center of the zone, was similar to that faced by the First Army on the opening day of the campaign. Liggett's plan called for the V Corps, still in the center, to thrust deep into the German lines, supported on the left by the I Corps, and on the right by the III Corps. This time, however, the objective was established at a more realistic distance of five miles. Liggett, with Pershing's blessing, met with Gen. Henri Gourard, commander of the French Fourth Army, on 25 October to coordinate their combined efforts. The French officer pointed out that his men would be unable to begin the offensive on the twenty-eighth, so the two generals agreed to begin the operation on 1 November.

The plan Liggett outlined for Gourard conflicted with the concept proposed by Foch earlier in the month. Foch had suggested that the two armies conduct a pincer movement aimed at conquering the Bois de Bourgogne, a wooded area in the vicinity of Grandpré at the extreme left of the American zone on the boundary separating the two armies. Liggett argued that a thrust by First Army to seize the Barricourt Heights in the center of the American sector would crack the Freya Line and put his forces in position to drive the Germans out of the Meuse-Argonne sector completely. Gourard saw no conflict in this maneuver, and it was agreed that Liggett would pursue it.

Because the American front was four miles narrower than it had been on 26 September, Liggett planned to employ only seven divisions in the line. All of these were experienced: the 77th, 78th, and 80th in Dickman's I Corps; the 2d and 89th in Summerall's V Corps; and the 5th and 90th in Hines's III Corps. Although the 78th, 89th, and 90th had remained in action during the period of reduced operations, Liggett thought they would be at sufficient strength to continue the offensive. The 5th Division moved into position on 27 October, and the remaining divisions joined the line within the next three days.[39]

The 3d Tank Brigade had not yet relieved the remnants of the 1st Brigade, so the only tanks available to support the operation were those of the 1st Provisional Tank Company. They were to be committed in support of the main effort, the V Corps thrust at the Barricourt Heights. As only fifteen tanks could be manned, it was decided to use all of them in the 2d Division's sector on a line between St. Georges and Landres-et-St. Georges—the same area in which they had last operated.[40]

At midnight on the night of 31 October–1 November, the Provisional Company moved out for assembly areas behind the 2d Division lines. Five tanks reached a position just south of Sommerance at 2:45 A.M. for duty with the 6th Marine Regiment, and the remaining ten arrived at the Bois de Romagne at 3:15. These platoons would support the 23d Infantry and 5th Marine Regiments. The tanks were met at the assembly areas by fuel trucks and topped off prior to moving to their departure points thirty minutes before H hour.[41]

A massive artillery preparation began at 3:30 A.M. Joining the First Army's batteries were a number of fourteen-inch naval guns mounted on railroad cars. These monster guns had been firing 1,400-pound projectiles up to twenty-five miles behind the German lines for several days. At 5:30 A.M. hundreds of whistles blew and the doughboys of the V Corps's spearhead divisions lurched from their trenches into the rolling November mists. This time, however, they employed a new twist: In the 2d Division sector the infantry were withdrawn 500 yards from the forward trenches, and a ten-minute barrage pounded the area just ahead of the line. This tactic was designed to eliminate German machine-gunners who, in the past, had moved close to the American positions to evade the artillery fire.[42]

As the tanks advanced ahead of the infantry, Barnard noted that "the Artillery work had been so thorough that there was very little wire to be cut and practically no machine-gun resistance except an isolated gun here and there which [the tanks] destroyed."[43]

All up and down the line the doughboys charged forward. By the time the 2d Division reached the St. Georges-Landres-et-St. Georges road, only six of Barnard's tanks were still running. The remainder had fallen prey to mechanical failure, tank traps, deep trenches, or shell holes. These six tanks, led by Lt. Floyd Callahan, continued forward, destroying German machine-gun positions in the high ground north of Landres-et-St. Georges. At 9 A.M. the first objective was secured and the tanks reorganized. Only four were found to be in fighting condition, and these were placed at the disposal of the commander of the 5th Marine Regiment for

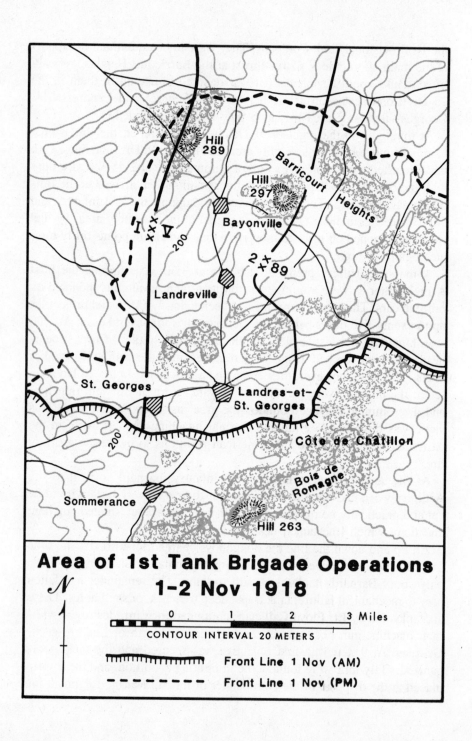

Hill
289

Hill
297

Barricourt Heights

I ✕✕ V
✕ ✕
200

Bayonville

2 ✕✕ 89

Landreville

St. Georges

Landres-et-
St. Georges

Côte de Châtillon

200

Bois de
Romagne

Sommerance

Hill 263

Area of 1st Tank Brigade Operations
1-2 Nov 1918

N

0   1   2   3 Miles

CONTOUR INTERVAL 20 METERS

⊥⊥⊥⊥⊥⊥⊥⊥⊥⊥⊥   Front Line 1 Nov (AM)

– – – – – – –   Front Line 1 Nov (PM)

the remainder of the day. At Landreville another tank went down with a broken track pin. The remaining three, led by Callahan, continued to press the attack. Several kilometers farther on they encountered a battery of four 77mm guns that they attacked from the flank, driving the crews from their posts. The enemy gunners fled into a nearby ravine, where they were captured by the advancing Marines along with a large stockpile of small arms and ammunition that had been abandoned there. Little further resistance was encountered the rest of the day. The tanks knocked out a lone machine-gun nest that had been "annoying" the infantry in the vicinity of the corps' objective, and some eighty prisoners were taken.[44]

Unlike the first day of the campaign, this time the First Army succeeded in reaching most of its objectives. Both the 2d and the 89th Divisions had advanced five miles into the German lines, scoring the day's best gains. Four German divisions, two of them considered to be first-class units, were overrun in the First Army zone. This forced the Germans into a general withdrawal toward the Meuse and the evacuation of the Bois de Loges and Champigneulle, two positions that had slowed the attack by the I Corps's 77th and 78th Divisions.[45] The speed of the German retreat posed a new problem: Even using trucks, Dickman's I Corps was unable to catch up with them.[46] It also effectively eliminated the 1st Brigade tankers from the operation. At the end of the day's operations on 1 November Callahan withdrew to Bayonville, where it was determined that none of the tanks were operable after the long road march. Attempts to get replacement tanks to the front from Varennes proved futile. On the second, seven tanks reached Landreville, leaving several others broken down along the road behind them. They were unable to catch up with the rapidly advancing line.[47]

On 6 November the AEF Tank Corps's brigades were redesignated to bring them in line with the system in use in the United States. The 1st Brigade became the 304th Tank Brigade, the 3d Brigade became the 306th Tank Brigade, and the newly organized 2d and 4th Brigades became the 305th and 307th Tank Brigades, respectively.[48]

The relief of the 304th Brigade was accomplished on 9 November, and its remaining forward elements returned to Bourg the following day. Although the 306th Brigade had sufficient manpower to crew all of the tanks still operational, the speed of the American advance made it impossible for the brigade to catch up with the line before the Armistice went into effect at 11 A.M. on the eleventh.[49]

The casualty toll in the 304th Brigade during the Meuse-Argonne campaign was three officers and sixteen enlisted men killed, and twenty-one officers and 131 enlisted men wounded. These casualties represented nearly 43 percent of the officers participating in the battle, and more than 21 percent of the enlisted men. Vehicle losses amounted to more than 123 percent. Despite the organizational maintenance problems encountered at battalion level, the fact that mechanics were able to continually recover, repair, and return tanks to action at the rate they did reflects well on their efforts. A total of eighteen tanks were destroyed by German fire.[50]

Where tanks were employed with seasoned infantrymen who had prior combat experience with tanks in support or had undergone training with tanks, they were particularly successful. This was especially true of operations on 4 October with the 1st Division, and on 1 November with the 2d Division. Unfortunately, these were the only experienced units with which the 304th Brigade had the opportunity to work. Maj. Gen. Robert L. Bullard, the III Corps (and later Second Army) commander, who had no tanks in support of his own divisions during the campaign, recalled after the war that he did not hear about tank operations until the campaign was well into its second week. The reports he heard were all favorable, but he was dismayed to learn that many of the doughboys "looked at their operation with great interest and admiration" but failed to adequately support them.[51] This fact was painfully apparent to the tankers, who bitterly complained that the infantry frequently held back and let the tanks do the dirty work for them.

A number of other observers spoke highly of the efficiency of the tanks, highlighting their ability to eliminate German machine-gun positions and strong points. Everyone, including the tankers themselves, bemoaned the fact that the machines were so mechanically unreliable, however.[52]

Despite these observations, the presence of tanks on the battlefield seems to have had little impact on the vast majority of the infantrymen they supported. With few exceptions, divisional histories are mute on the subject of the tankers' contributions during the Meuse-Argonne and St. Mihiel campaigns. Nevertheless, the losses suffered by the 1st/304th Tank Brigade and the relatively high percentage of valor awards given to the unit's personnel bear testimony to the severity of the combat in which they engaged.

There is little doubt, however, that the presence of the tanks on the battlefield had a significant impact on German morale. German prisoners reported that the mere sight of a tank sent a ripple of terror through their ranks. It was probably with this in mind that Rockenbach concluded his report on the 304th Brigade's operations in the Meuse-Argonne:

> Of course, now that it is over, to the survivors it was like the bear story. The Tank said "Booh!" and the Boche said "Booh." The Tank rattled its tread, the Boche unable to make such a noise as that, . . . he beat it.[53]

Notes

1. Coffman, *War to End All Wars,* 321.
2. Maj. Sereno E. Brett, "Operations Report of the 1st (304th) Tank Brigade from Noon, September 26th, 1918, to November 10, 1918," included as Appendix 6 to Rockenbach, "Tank Corps Operations," 3.
3. Rockenbach, "Tank Corps Operations," 31–32.
4. Snyder, "Company 'C' 344th Battalion, 2d Platoon, Argonne Sector"; 2d Lt. H. A. Wood, "3rd Platoon Company C 344th Battalion (304th Brigade), Tank Corps, Argonne Sector," undated, Patton Military Papers, Box 47, Personal Experience Reports of Tank Operations–1918, Patton Collection.
5. Patton, "Citation of Members of First Brigade, Tank Corps"; General Order no. 24, GHQ–Tank Corps, AEF.
6. Patton, "Citation of Members of First Brigade, Tank Corps."
7. McClure, "Personal Experience Report."
8. General Order no. 24, GHQ–Tank Corps, AEF.
9. Ibid.
10. Ibid.
11. Compton, "345th Battalion War Diary."
12. Patton, "304th Brigade History," 39.
13. Compton, "345th Battalion War Diary"; General Order no. 24, GHQ–Tank Corps, AEF; Patton, "Citation of Members of First Brigade, Tank Corps."
14. Patton, "304th Brigade History," 40; Coffman, *War to End All Wars,* 323.
15. Stallings, *The Doughboys,* 266–77.
16. Patton, "Citation of Members of First Brigade, Tank Corps."
17. Patton, "304th Brigade History," 41.
18. Brett, "Appendix 6, Tank Corps Operations," 5; Compton, "345th Battalion War Diary."
19. Ibid.
20. Rockenbach, "Tank Corps Operations," 32–34.
21. Wahl, "Appendix 7, Tank Corps Operations," 11–12.
22. Coffman, *War to End All Wars,* 327.
23. Patton, "304th Brigade History," 43; Brett, "Appendix 6, Tank Corps Operations," 6.
24. Patton, "304th Brigade History."
25. Atwood, "Personal Experience Report."
26. Patton, "304th Brigade History," 44.
27. Atwood, "Personal Experience Report."

28. Coffman, *War to End All Wars,* 329.
29. Ibid.
30. Patton, "304th Brigade History," 44. The lone missing tank was believed to have been lost in the Aire River but was recovered after the Armistice (Rockenbach, "Tank Corps Operations," 26).
31. Special Orders no. 290, General Headquarters, AEF, 17 Oct. 1918, Patton Chronological Files, Box 11, 1–18 Oct. 1918, Patton Collection.
32. Capt. John R. Johnson, "Report of Investigation French Renault Tanks with 1st Army, American E. F., Argonne Sector, October 1918," 25 Oct. 1918, included as enclosure to Maj. R. E. Carlson, "Memorandum on the Development of Tanks," 16 Mar. 1921, USAMHI.
33. Brett, "Appendix 6, Tank Corps Operations," 6–8.
34. As quoted in Blumenson, *Patton Papers,* 631.
35. Special Orders no. 518, Headquarters, 1st Army, AEF, 29 Oct. 1918, Modern Military Records Division, Record Group 165, Entry 310, Box 446, File 66-12.8, NA.
36. Coffman, *War to End All Wars,* 332.
37. Ibid., 332–33.
38. Questionnaire, Col. Charles H. Phillips, 306th Tank Brigade, 316th Repair and Salvage Company, World War I Survey Collection, Archives, USAMHI.
39. Coffman, *War to End All Wars,* 343–44.
40. Patton, "304th Brigade History," 45.
41. Capt. Courtney H. Barnard, "Report of Operations 1st Provisional Tank Co., 1st Brigade Tanks, 1 November 1918," undated, Modern Military Records Division, Record Group 165, Entry 310, Box 446, File 66-33.6, NA.
42. Coffman, *War to End All Wars,* 345.
43. Barnard, "1st Provisional Tank Co. Report of Operations."
44. Ibid.; Patton, "304th Brigade History," 45.
45. Coffman, *War to End All Wars,* 345–46.
46. Ibid., 346.
47. Barnard, "1st Provisional Tank Co. Report of Operations."
48. General Orders no. 21, General Headquarters, Tank Corps, AEF, 6 Nov. 1918, Modern Military Records Division, Record Group 120, Entry 1296, NA.
49. Brett, "Appendix 6, Tank Corps Operations," 9.
50. Ibid., 11.
51. Maj. Gen. Robert L. Bullard, *Personalities and Reminiscences of the War* (Garden City, N.Y.: Doubleday, Page & Co., 1925), 276.
52. Lt. Gen. Hunter Liggett, *AEF: Ten Years Ago in France* (New York: Dodd, Mead & Co., 1928), 325. Several divisional histories briefly mention the efficiency of tanks on the battlefield, especially in the Meuse-Argonne and St. Quentin Canal battles. Among these are Kenamore's 35th Division his-

tory, Maj. Gen. John F. O'Ryan's history of the 27th Division, and the Society of the First Division's unofficial history. None of the divisional summary histories prepared by the American Battle Monuments Commission do more than acknowledge the Tank Corps's presence on the battlefield.

53. Rockenbach, "Tank Corps Operations," 35.

CHAPTER IX

The Heavy Tanks with the British Expeditionary Force

By mid-September 1918, the officers and men of the 301st Heavy Tank Battalion were growing restless. For more than two weeks they had done little more than perform maintenance on their newly acquired tanks, hoping all the while that orders would come down sending them "over the top." Second Lt. John A. Logan, a tank commander in Company C, recalled that one private "expressed the wish of the entire battalion when he said, 'Show me the way to the war that we have come so far to take part in.'"[1]

The long-awaited word finally arrived on 19 September, when the commander of the British 4th Tank Brigade met with Maj. Ralph I. Sasse and the 301st's company commanders at the battalion's headquarters at Achiet-le-Grand. Their discussion focused on plans for the brigade's support of the British Fourth Army's upcoming assault on the Hindenburg Line in the vicinity of the St. Quentin Canal.

Sasse, who had been the battalion tactical officer, succeeded Lt. Col. Henry E. Mitchell as battalion commander on 9 September, when Mitchell left to observe the 1st Tank Brigade's participation in the St. Mihiel offensive prior to assuming command of the soon-to-be-activated 2d Tank Brigade, which would become the 301st's parent organization. The move was well received in the 301st, where the hard-driving Sasse had earned the tankers' respect and where Mitchell had been held in "low esteem."[2]

The plan briefed to the Americans called for the brigade to support a 29 September attack by the Fourth Army's Australian Corps. This assault was to be the last of four general attacks slated to begin on 26 September as part of Field Marshal Sir Douglas Haig's proposed pincerlike movement aimed at forcing the Germans out of France.

Gen. Henry Rawlinson, the Fourth Army commander, whose force included the Australian Corps and American II Corps, decided to give operational control of the American troops to Lt. Gen. Sir John Monash, commander of the Australian Corps. Monash, recognizing the battle-weary condition of his own soldiers, elected to employ the II Corps's 27th and 30th Divisions in the initial assault. In Monash's opinion, the mission was a simple one. The terrain in his sector was open and rolling—ideal for the employment of heavy tanks. The key terrain feature was the St. Quentin Canal, a remnant of the Napoleonic era. The Germans had placed their main line of resistance behind this obstacle wherever possible, but in Monash's zone the canal posed little problem as it went into an underground tunnel spanning about 6,000 of the 7,000 yards on the Australian front. Monash proposed that his force avoid a crossing of the canal, directing the main effort at the triple layer of trenches in the tunnel area instead. Once these had been breached, he wanted the Americans to assault the Beaurevoir Line, then pass through his Australian divisions, which would exploit the breakthrough. Rawlinson was less optimistic, however. He overruled Monash, limiting his objective to a penetration of the third line in the Hindenburg system. Furthermore, he did not wish to dismiss a crossing of the canal from consideration, although he assigned this mission to the adjacent British corps operating south of Monash.[3]

The 4th Tank Brigade planned a simple "two up, one back" scheme of maneuver for the operation, with Sasse's 301st Battalion supporting the 27th Division with forty tanks in the left (north) half of the Australian Corps's zone, and the British 1st Tank Battalion supporting the 30th Division's attack to the south. The British 4th Tank Battalion and seven 301st Battalion tanks would be held in corps reserve. Maintenance and logistical support were to be provided by the British 4th Tank Supply Company.[4]

Armed with this information, Sasse and the battalion spent the remainder of the nineteenth and the twentieth preparing the tanks for rail shipment to Equancourt. First Lt. Carleton Reynell, the 2d Tank Brigade reconnaissance officer, and a Lieutenant McClusky, the 301st's reconnaissance officer, departed ahead of the rest of the battalion to scout the area around Manancourt and select the location for a "tankodrome."[5]

The 301st had been equipped with two basic tank types by the British: the Mark V Star and the Mark V. The Mark V was considerably smaller than the Star tank, measuring some six feet shorter (twenty-six feet, five inches compared to thirty-two feet, five inches), weighing in five tons lighter, and mounting one less machine gun. Both tanks came in "male" and "female" versions. The male tanks were armed with a mix of six-pounder guns and Hotchkiss 8mm machine guns, while the female tanks mounted machine guns only. A third derivative of both types, called a composite or hermaphrodite, was equipped with a single six-pounder in lieu of one of the Hotchkiss machine guns.[6] These vehicle types were employed in varying mixes by the 301st's three companies. Company A had nine Star males, two Star females, and four Star composites. Company B was assigned seven Star males, two Star females, three Mark V males, one Mark V female, and three Mark V composites. Company C received seven Mark V males and nine Mark V composites. In addition, all of the shorter Mark V models were equipped with fascines, cumbersome devices about eight feet in height made of heavy timbers and metal facings. These were strapped to the front of the vehicle and were used to assist in the crossing of wide, deep trenches.[7]

On 21 and 22 September the battalion rail-loaded its forty-seven heavy tanks, plus a Renault light tank provided for use by the headquarters, onto four trains (two each day) for the move to Equancourt. In each case, off-loading and movement to the "tankodrome" near Manancourt were accomplished without incident after dark, although Lieutenant Reynell recalled that intermittent artillery fire was received. One shell landed within a few yards of one of the battalion's trucks, but there were no casualties.[8]

Sasse again met with the 4th Tank Brigade's commander on the twenty-third and was instructed to meet with the 27th Division commander, Maj. Gen. John F. O'Ryan, to work out detailed plans for the operation. Sasse and O'Ryan conferred on the night of the twenty-fourth. The two agreed that the main part of the 301st's effort should be made in support of the 54th Brigade, which would lead the division's attack. To accomplish this mission, Sasse assigned Company C with fifteen tanks to support the brigade's 107th Infantry Regiment. Company A, also with fifteen tanks, would support the 108th Infantry Regiment. The battalion's remaining ten tanks, all from Company B, were tasked with supporting the 53d Brigade's 105th Infantry Regiment, which would follow the 107th.[9]

Given this guidance, Sasse returned to Manancourt, where he worked out a more detailed scheme of maneuver. Following approved doctrine for

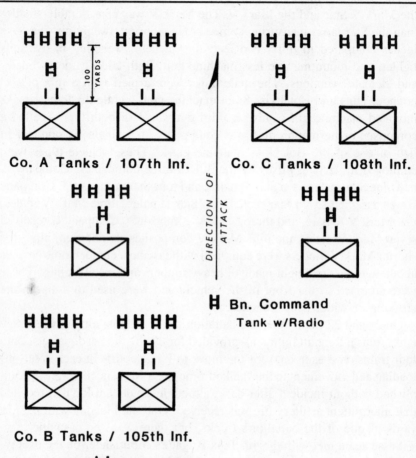

Co. A Tanks / 107th Inf.

Co. C Tanks / 108th Inf.

DIRECTION OF ATTACK

100 YARDS

H Bn. Command
Tank w/Radio

Co. B Tanks / 105th Inf.

301st Tank Battalion

SCHEME OF MANEUVER

FOR OPERATIONS WITH 27TH INFANTRY DIVISION

29 SEPTEMBER 1918

the employment of heavy tanks, he instructed his company commanders to echelon their tanks by assigning a platoon to each of the regimental infantry battalions. The regiments would conduct the attack in a "V" formation, with two battalions forward and one in reserve. The tanks were to advance about 100 yards in front of each battalion, clearing wire obstacles and destroying German machine-gun and artillery emplacements.[10]

While Sasse was working out his scheme of maneuver, battalion and company reconnaissance officers began a detailed study of the terrain. Because surprise was important, physical reconnoitering was limited. Working mostly from aerial photographs, the reconnaissance officers identified obstacles, trenches, and gun emplacements, recording the information on a special map maintained by Lieutenant McClusky. Locations for the battalion's "lying-up point" (forward assembly area), "starting point" (line of departure), and rallying point were provided by the 4th Tank Brigade. The battalion's reconnaissance officers selected routes and prepared overlays depicting all of this information for the battalion and company commanders, platoon leaders, and tank commanders.[11]

The 27th Division's sector included the towns of Le Catelet, Gouy, and Bony, as well as several strongly fortified farms. Lieutenant Reynell noted during a trip to the front that the Germans, recognizing that the area in which the St. Quentin Canal went underground was particularly inviting for tanks, "fortified it as strongly as possible." In addition to the trench system, a formidable obstacle itself, the Germans had dredged up "slag heaps" and had "wonderfully fortified [the area] with acres of wire and a great amount of artillery."[12]

On 25 September Sasse, accompanied by all of his officers down to platoon level, attended a conference at 27th Division headquarters. When the meeting ended, the tank officers conferred with the commander of the 54th Brigade, who approved Sasse's scheme of maneuver. Sasse was concerned by the division's lack of experience with tanks and later observed that the infantry officers "did not seem to grasp the idea of Tanks co-operating with Infantry." Because of this, and because there was not enough time to conduct rehearsals for the operation, Sasse instructed his company commanders and platoon leaders to meet daily with the infantry unit commanders to discuss detailed plans and to work out mutually agreed-upon means of coordination to be employed during the battle.[13]

While the tankers and doughboys made their final preparations for "Z" day, a major flaw in Monash's plan was discovered: The Australian Corps commander had based his operational planning on the assumption that the attack would begin from positions then in German hands. The projected start line roughly followed the outpost line of the Hindenburg system. This posed no problem for the 30th Division, as the Australians had forced the Germans back far enough to make it relatively easy for them to secure the jump-off line. The 27th Division was less fortunate. The two British divisions it relieved had made no effort to take the position. This meant the 27th would have to fight its way about 1,000 yards and capture the German trenches. The mission was assigned to the 53d Brigade's 106th Infantry Regiment.

At 5:30 on the morning of 27 September the 106th launched an assault aimed at gaining the required ground. A handful of British tanks were provided to support the maneuver but proved to be of little help. Bitter fighting raged throughout the day in the vicinity of three German strong points: the Knoll, Guillemont Farm, and Quennemont Farm. Three times a reinforced battalion seized the Knoll, only to be thrown back after each attempt to secure it. Repeated German counterattacks resulted in the 106th's withdrawal to its own lines by dusk, leaving an unknown number of casualties and stragglers isolated in no-man's-land and behind the German line.

The failure of the division to secure the starting line was especially problematical because, unlike the battle of Cambrai, in which no artillery preparation had been employed, thus enhancing the element of surprise generated by tanks attacking en masse, the Fourth Army had laid on a massive sixty-hour preparation culminating in a rolling barrage set to begin at the start line at H hour. All of the artillery firing tables had been completed, and there was not enough time to recalculate them before the attack kicked off. O'Ryan requested a one-day delay in the operation to give him time to mount another attack to seize the jump-off positions, and Monash dutifully passed the request along. Rawlinson denied it, however, pointing out that three armies were involved and that he could not alter the plan to help out one division. Instead, he suggested that O'Ryan launch his attack one hour earlier in hopes the 27th's doughboys could cover the 1,000 yards before H hour.[14] This decision would have a dramatic impact on the tankers, who were already experiencing difficulties dealing with their infantry counterparts.

The final forty-eight hours before the attack were hectic ones for the 301st's reconnaissance officers. To ensure surprise, all movement behind the front was conducted under cover of darkness. This meant routes had to be marked with white tape and careful coordination made with other

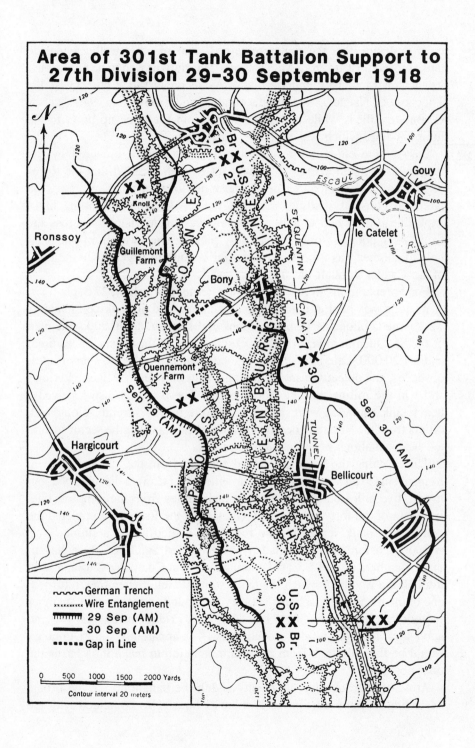

Area of 301st Tank Battalion Support to 27th Division 29–30 September 1918

N

Br. 18

XX 27

U.S. LINE

Escaut

Gouy

XX

The Knoll

le Catelet

Ronssoy

ST. QUENTIN

Guillemont Farm

Bony

CANAL

XX 30

ZONE

Quennemont Farm

Sep 29 (AM) XX

HINDENBURG

Sep 30 (AM)

OUTPOST

Hargicourt

TUNNEL

Bellicourt

U.S. Br. 30 XX 46

XX

| German Trench |
| Wire Entanglement |
| 29 Sep (AM) |
| 30 Sep (AM) |
| Gap in Line |

0   500   1000   1500   2000 Yards

Contour interval 20 meters

units to ensure that the routes remained clear of other traffic when the tanks moved forward from Manancourt. This duty proved to be hazardous: Lieutenant McClusky and Company A's reconnaissance officer, 1st Lt. Theodore C. Naedele, were both wounded, and a draftsman, Sgt. Charles C. Butler, killed, while taping the route from the lying-up point to the jump-off line, which was still in enemy hands.[15]

Last-minute efforts to reconnoiter the sector were hindered by the lengthy artillery preparation. Lieutenant Reynell recalled that the towns of Ronssoy and Haricourt, located near the forward positions, were the target of constant shelling, and that the smoke from exploding high-explosive rounds and clouds of mustard gas made visibility across the front "very poor . . . only by carefully picking one's way from one shell hole to another was it possible to get far enough forward to see the town of Le Catelet through the field glasses."[16]

Final preparations were made on the afternoon of Friday, 27 September. A meeting was held at battalion headquarters, and all tank commanders were issued oblique aerial photographs of the sector, a 1:10,000 scale map of the St. Quentin area, a 1:100,000 scale map of the Valenciennes area, and a 1:20,000 scale map of the Wiancourt area, marked to show routes, obstacles, and objectives and including a table showing the lines and times shifts in the rolling barrage would occur. The battalion also received a heavy tank equipped with radio gear on the twenty-seventh, along with orders from the British to deploy the tank in the center of the sector behind the lead battalions of the assault regiments.[17]

That evening the battalion moved forward from the "tankodrome" at Manancourt to an assembly area near Villers Fan Con—a distance of about twenty-five miles but still more than eight miles from the jump-off line. Forty-six of the forty-seven heavy tanks, plus the Renault, completed the march. One tank veered off the route and bellied up on a stump.[18]

Shortly before the long road march, recalled Sgt. Carl Rosenhagen, three members of 2d Lt. Earl Dunning's crew in Company C took a walk through the woods in which the tanks were hidden. When they returned, one of the men carried a black cat they had found. Another crew member wryly observed that finding a black cat on a Friday must surely portend good luck for the unit's thirteenth crew.[19] It apparently did: Dunning's would be the only tank from the entire battalion to make it as far as the final objective.

At 4 P.M. on Saturday the twenty-eighth the battalion received orders advancing its jump-off time to H-minus-1 hour. This coincided with the earlier jump-off time for the assault regiments, which had to make up the ground the 106th Infantry Regiment had failed to take the day before.[20]

The 4th Tank Brigade had established a supply dump near Villers Fan Con, and all of the tanks were able to top off with fuel and take on a full load of oil and grease.[21] This was no easy task. Fuel was issued in two-gallon cans that had to be passed from man to man and tank to tank, then carefully poured into the armored fuel tanks, all while sporadic German artillery fire rained down on the area.[22]

At 10 P.M. the battalion moved forward to the lying-up point, a distance of about 5,500 yards. All but two of the tanks made it: The little Renault broke its track, and one of the heavies stripped its gears. The move was especially difficult because the Germans shelled the area with a mixture of high-explosive and gas rounds, wounding several men and slightly gassing one crew.[23] Recalling the experience later, Sergeant Rosenhagen observed:

> It was sure a hardship driving with a mask on. The manual labor required to drive a tank over rough territory with the heat of the motor and confined space inside made you perspire profusely. With both hands and feet working, it was next to impossible to keep the nose clips [of the gas mask] from slipping off, and the gas burned terribly around your face and eyes; and to breathe through the mouthpiece and see where you were going was really brutal punishment. It was impossible for the driver not to breathe in part of the gas.[24]

Captain Ralph deP. Clarke, the Company C commander, said the approach march to the lying-up point was particularly demanding on officers, trying their "very souls" as they struggled on foot in advance of their tanks over sunken roads and muddy fields, through barbed-wire entanglements, and across shell holes and abandoned trenches. Company commanders, said Clarke, had their nerves "strained to the point of breaking" as they groped through the dark a hundred yards in front of their units, the "fear of losing the way and not arriving in time, or a chance hit [by artillery] putting part of" the command out of commission constantly preying on their minds. Platoon leaders hustled up and down the line of vehicles in an effort to keep the column moving smoothly, and individual tank commanders preceded their vehicles on foot, communicating instructions to their drivers by frantically waving a glowing cigarette.[25]

Lieutenant McClusky, the battalion reconnaissance officer, who had been wounded earlier that night while laying tape marking the route to the start line, relayed a message to Lieutenant Reynell that Sergeant Crace and Corporal Pfaff, two headquarters draftsmen, would be on hand to guide the A Company tanks forward to Ronssoy. The town was being heavily shelled, and Capt. Kit R. Varney, the company commander, dismounted from his tank to personally lead his column forward. A dense fog settled over the battlefield, limiting visibility, and as dawn and H hour approached, Varney continued forward on foot, leading his tanks several hundred yards past Ronssoy into no-man's-land before being killed by a burst of German machine-gun fire.[26]

As Company A passed the old British front line, it encountered an anti-tank mine field. The British had failed to notify Sasse that the mine field, laid prior to the German spring offensive, was there. The Americans were taken completely by surprise, and the tanks commanded by 2d Lts. Fred C. Jones and Joseph J. Gutkowski were destroyed. The tanks commanded by 1st Lt. James Taylor and 2d Lts. Harold R. Dean and Edward Colburn were forced to turn back because of mechanical problems shortly after crossing the forward infantry positions.[27]

But Company A at least had the support of the 108th Regiment, whose doughboys followed the tanks forward to the start line. Company C was less fortunate. Shortly before 5 A.M. on the twenty-ninth Sasse arrived at the 107th Regiment's command post, where he was informed by the regimental commander that the 107th would not be moving out until 5:55 A.M.—the originally scheduled H hour. Rather than wait for the doughboys, Sasse adhered to his orders and waved Company C forward. His decision resulted in Company C's fighting without infantry support throughout the day, as the 107th's doughboys were never able to close the gap created by the tankers' one-hour head start.[28]

Despite the fog, German artillery fire proved deadly accurate. Three tanks from Company A were knocked out by direct hits prior to reaching the start line. In Company C's sector, Captain Clarke was forced to abandon his tank when the driver and other crew members became severely disoriented after experiencing several near misses as they approached the jump-off position.[29]

Cpl. Albert F. Neil and Pvt. Robert F. Wisher helped drag wounded crew members from their shell-shattered A Company tank, then dismounted a pair of machine guns, pouring fire into the forward German trenches in the vicinity of the start line until they were forced to abandon the guns. Returning to their tank, the pair secured rifles and hand grenades, then joined forces with advancing doughboys from the 108th Infantry Regiment and organized an assault on the German positions. The attack was successful, but they were forced to fall back after casualties depleted their ranks. Unable to make contact with American units that had passed them by, Neil and Wisher linked up with advancing Australian troops and spent the rest of the day fighting with them.[30]

Second Lt. Le Roy B. Mitchell's tank was another of the A Company tanks knocked out en route to the start line. One of Mitchell's gunners, Pvt. Wayne Hughes, recalled that the crew, despite suffering ill effects from the gas employed against them during the approach march, continued forward until a heavy-caliber shell wrecked the vehicle. Two crewmen were killed, and the remainder were wounded by the blast. Hughes suffered shrapnel wounds in the legs, hands, head, and face but managed to drag himself free.[31]

Dawn and H hour arrived as A Company's remaining seven tanks cleared the start line, followed closely by the lead battalions of the 108th Regiment. The rolling barrage began on schedule, adding a mixture of smoke rounds to conceal the tanks' movement. While the smoke proved helpful early in the attack, the wind soon shifted, blowing the smoke back toward friendly lines. The swirling fog and smoke made visibility virtually impossible beyond five yards, and tank commanders tossed aside their carefully prepared maps, relying instead on compasses to keep them on course. Many of the compasses proved to be defective, however, and soon tanks were scattered all over the battlefield.[32]

One of those who became lost was 2d Lt. John A. Logan, a C Company tank commander. Logan and his crew cleared the jump-off line on the right flank of the formation. As the tanks advanced, Logan recalled, his own vehicle

rocked like a ship, for Fritz had put down a barrage and shells were falling very close. We could see the other tanks mopping up the trenches to the left of us and four tanks to the right, which we thought were from our platoon. After fighting alongside of these four tanks for a while, we discovered them to be from A Company, so [we] turned to the left to join our own Company. . . . We had just cleared up a trench of Jerries when our engine went dead. We thought that an armor piercing bullet had fixed us, but after a few minutes' cranking, we managed to get a couple of cylinders coughing so that we were able to get the tank into the trench so as to repair the damage. Suddenly I heard a yell and turned to see a mass of flames shooting from the engine. The whole tank seemed to be on fire. The driver and I tried to get out through the top but the door stuck, so back thru the flames we went. I tried to put the fire out with Pyrene but the fumes made me groggy almost instantly and but for the fact that Sgt. Hoberts dragged me out I would have surely been badly burned. We were scarcely outside when a German aeroplane flying low dropped a couple of flares. Just then business began to pick up, for the shells began to fall all about us.[33]

Sasse reported later that a number of crews recalled similar experiences. It was apparent, given the volume and accuracy of the Germans' artillery fire, that a system combining both ground and aerial observers was being used to pinpoint the location of the tanks for the artillery batteries.[34]

A German machine-gunner spotted Logan's crew in the trench and began raking the Americans with fire. One of Logan's gunners was able to dismount a pair of machine guns from the tank, and the crew fought back, covering the advance of supporting doughboys, who helped them clear the position.[35]

As the 301st pressed the attack on the first series of trenches in the Hindenburg Line, Sasse and Reynell halted the radio-equipped headquarters tank in a shallow valley from which they could observe the tankers' initial progress. Visibility was hindered by the smoke and fog, and the Germans made matters worse by mixing gas rounds in with their counterbarrage. Reynell and Sasse crawled forward in an effort to keep the advancing tanks in sight, only to become separated in the swirling mist. They maintained contact by yelling to each other, finally reuniting back at the radio tank. A momentary break in the fog and smoke permitted them a glimpse of several tanks on the ridge to their front. Sasse ordered the crew to prepare to move forward, but the fog quickly closed in, and he decided to remain in the position until conditions improved.[36]

A Company's ranks were further thinned during the assault on the first trench in the Hindenburg system when the tanks commanded by 1st Lt. Elliott P. Frost and 2d Lt. A. B. Maury ditched in the German position. Frost waved down Second Lieutenant Adams, and he and his men joined Adams's crew, continuing to advance until the tank was destroyed by a direct artillery hit. The crewman next to Frost was killed, and several other men were wounded by the blast. With Adams's tank out of action,

Frost led the remainder of his crew back to his own vehicle, where they worked for two days and nights without support, finally succeeding in freeing the tank from the trench and rejoining the battalion.[37]

First Lt. John R. Webb's tank ditched in a shell hole just forward of the Germans' front-line trench. The crew quickly dismounted and began trying to dig the vehicle out. A machine gun located about thirty yards away opened up on them, however, sending the men scurrying for cover. Unable to bring any of the tank's machine guns to bear on the German position, Webb crawled forward and killed the crew with his pistol. Webb's action enabled his men to get back to work. They subsequently freed the tank and continued to advance. A nearby tank commanded by 2d Lt. Clarence J. Hartt stalled when shell fragments damaged its track. Heavy German machine-gun fire kept the crew from dismounting to attempt repairs. While the gunners returned fire, Cpl. Frank W. Osmond crawled out of the back of the vehicle and used a rifle to pick off German machine-gunners. With the enemy fire suppressed, the crew was able to repair the damaged track and continue to advance in support of Webb.[38]

Fighting in the trench system was particularly bitter. The smoke and fog made visibility difficult for the three remaining A Company tanks, and they were frequently forced to engage German machine-gun positions at extremely close range, often rolling over the Germans and grinding out the opposition with their tracks. The tank commanded by 2d Lt. Morton B. Hillsley was damaged by artillery fire and forced to turn back, and Webb and Hartt ran out of ammunition before reaching the canal tunnel, bringing A Company's participation in the attack to a close.[39]

Company C's attack fared little better. Ten of the company's fifteen tanks cleared the start line but were forced to advance without infantry support, as the 107th Regiment was still in its trenches more than a thousand yards to the rear at zero hour on the twenty-ninth. Almost from the moment they crossed the start line, the C Company tanks came under intense artillery and machine-gun fire. Second Lieutenant O'Kane and one of his crewmen were killed shortly after crossing the start line when their tank was destroyed by a direct artillery hit. Not far away, Second Lieutenant Shanks, unable to see clearly in the smoke and fog, ran his tank into a ditch.[40]

Second Lt. Samuel McKay, who had earned the Medal of Honor during enlisted service in the Marine Corps, was wounded twice by shell fragments while manning his tank's observation tower during the first minutes of the attack. Overcome by pain and unable to perform effectively, he decided to leave the vehicle in command of one of his NCOs but was shot and killed while climbing out. The crew continued to advance, but the men were forced to abandon the tank short of the German front-line trench when it was hit by artillery fire, seriously wounding several crew members.[41]

Two more C Company tanks were ditched in outpost trenches, and the tank commanded by Second Lieutenant Potter was destroyed by a direct artillery hit that killed six enlisted crewmen and seriously wounded the rest of the tank's occupants, including the platoon leader, First Lieutenant Rosborough, and his two orderlies. The survivors hastily evacuated the burning vehicle, seeking shelter in a nearby shell hole. Rosborough, though wounded and burned himself, dressed the wounds of the others before starting the painful trek back to the 107th Regiment in search of assistance. He left one of his orderlies, Sgt. Frank J. Williams, in charge, as Lieutenant Potter was in serious condition. Williams and the other wounded crewmen dismounted several Hotchkiss machine guns and set up a hasty defense, managing to hold out until lead elements of the 107th arrived to relieve them. Rosborough was spotted by several Germans manning an outpost position as he crawled toward the rear. Unable to resist, he was captured. Two of the Germans escorted him through a trench, but he managed to escape by pushing one guard down and hitting the other in the eye. He scrambled over the side of the trench and crawled away, spending the night and most of the next day hiding out under a wrecked tank.[42]

The only C Company tanks to make the first trench in the Hindenburg system were those commanded by Second Lieutenants Dunning and Little. Little's tank soon experienced mechanical problems and was forced to turn back. It was later destroyed by artillery fire while returning to the rallying point. Dunning, however, was able to press the attack, and his was the only 301st Battalion tank to cross the canal tunnel and reach the objective area near Le Catelet.[43]

Later, recalling the events of that harrowing morning, Sgt. Carl Rosenhagen wrote:

> I had turned the driving over to Sergeant Barnard . . . and was in the observation tower standing on a platform over the transmission taking care of anything that might happen and picking out targets for the gunners. Before we came to the antitank trench, we ran into about 20 machine guns corralled together, firing against the right side of the tank. The whole inside

of the tank seemed to be on fire from the sparks of the armor-piercing bullets around the gun slots; and through the other slots our gunners had to see their targets. It was so intense that one gunner laid back away from his gun, and I jumped down and took his place. . . . [He] looked up from the gun he was on, and his face was a mass of blood. Seeing we could not cope with these machine guns, I pounded on the motor cowling, which was how we attracted the driver's attention, and put up my fist [to signal] a left turn. He must have had his window or flap wide open, because he turned to the right, directly into those machine guns, and was badly hit. They hollered for first aid, and I jumped to him and said, "For God's sake, Barney, keep on driving until we get through this mess."[44]

Barnard stayed at his post, turning the tank around so the gunners could bring machine-gun and six-pounder fire on the position, knocking it out. Rosenhagen then relieved Barnard as driver and continued the advance. As they approached the Germans' front-line trench, Rosenhagen halted and removed the nuts on the chains holding the fascine to the front of the tank. He then drove forward, but the chains snagged, and the fascine refused to drop. Rosenhagen backed up a short distance and approached the trench at an angle. The vehicle bogged down, however, and it took several minutes of back-and-forth movement to break down the sides of the trench enough to permit the tank to extricate itself.

Smoke and fog continued to obscure the battlefield, but Dunning's crew pressed on. Arriving on the outskirts of Le Catelet, Rosenhagen stopped the tank near a crossroads, and the crewmen opened the hatches to let in some fresh air. While Rosenhagen and Lieutenant Dunning were looking over a map spread out between them in the tank's gloomy interior, the sergeant spotted several men running toward the vehicle. Thinking they might be friendly, Rosenhagen grabbed Dunning as the officer reached for the nearest machine gun, saying, "Don't fire—they are Englishmen." But he was mistaken. Dunning, realizing the approaching troops were German, attempted to fire the gun, but it jammed. One of the Germans shoved a rifle through the front hatch on Dunning's side and shot a piece out of the lieutenant's nostril. Rosenhagen quickly "dropped into gear and gave the tank a leap, as Dunning used his Colt Automatic." The crew responded to the attack by closing all the tank's hatches, and "we had a little commotion for a few minutes, swinging the tank around to get the Germans off the side machine guns so that our six-pounder could get to them. [But] our six-pounder crew was really good and made short work of them."[45]

Surrounded by Germans and without infantry support, Dunning ordered Rosenhagen to head back toward friendly lines in search of the supporting infantry. As the tank headed toward the rear, a German shell hit the back of the vehicle, breaking the main line from the water cooler and sending a spray of hot steam throughout the crew compartment. Rosenhagen ordered the crew to open all of the ports and continued driving in a southwesterly direction until a second shell landed squarely atop the tank:

> I believe the motor casing saved us; we had a big plop inside, and we caught fire. The whole tank seemed [to be in flames]. I climbed through Lieutenant Dunning's seat, the only way I could get out, and came along the motor, by the six-pounder, over the transmission; then I saw the lower door open on the right side and I slid out. In the smoke and fog I could see two of our boys running away. I hollered to them and they came back. Lieutenant Dunning got his face burned getting through and out of the tank. I was lucky, as I kept my head buried in my arms and only lost some excess hair off my head.
>
> The four of us ran for some small trenches we could see ahead. German machine guns were firing a barrage, and we laid there concealed by the fog and smoke, seeing only the fire coming out the barrels of the machine guns. We crawled on, and suddenly, about 11 o'clock that morning, the fog and smoke screen started to lift. We jumped into some shell holes. We could then make out our tank about 800 or 900 feet away. Lieutenant Dunning and I were in the same shell hole, not very big. Didn't see the other two boys until later that night. Our tank was burning and blowing up all afternoon with smoke belching out. German planes were flying right over our heads, but they must not have seen us lying below. I think the Germans figured we never got out of the tank.[46]

Dunning and Rosenhagen's ordeal was far from over, however. Under cover of darkness, the pair began "hours and hours of patient, agonizing crawling and wriggling and waiting," finally reaching friendly lines. They reported to the battalion the next night, "sick from gas and exposure, exhausted, soaked, [and] torn to ribbons by the wire." They were the only ones to escape; the other seven crewmen were captured.[47]

Company B, leading the assault battalions of the 105th Infantry Regiment with ten tanks, was just as luckless. Only eight tanks reached the start line, two having turned back with mechanical problems. Of the eight that started the attack, only one, Second Lieutenant Ellingwood's, reached the Hindenburg Line. Ellingwood was able to press the attack as far as the third trench line before being forced to turn back with mechanical problems. Two of the remaining tanks also experienced mechanical failures, and the other five were knocked out by artillery fire.[48]

The experience of 2d Lt. E. F. Kusener was representative of those who led the 105th toward the Hindenburg Line. After clearing the start line, Kusener quickly became disoriented in the smoke and fog. He dismounted and led his tank forward several hundred yards in the face of withering machine-gun bursts and devastating artillery fire. One round scored a direct hit on the tank, killing three crewmen and wounding four. Kusener rushed back to the vehicle, where he was able to pull the wounded from the burning tank, depositing them in a shell hole and carrying the two most critically injured men back to an aid station for treatment.[49]

By 10 A.M. on the twenty-ninth, the fog had begun to lift, and the creeping barrage had progressed far enough to allow the smoke to dissipate, permitting Sasse and Reynell to move forward in the radio tank. The 301st's commander could plainly see that the attack had not gone well for his tanks. Several were burning brightly on the battlefield, and the scattered remnants of their crews began trickling in. Sasse ordered Reynell to set up behind a hedge, then moved off on foot to assess the situation. Finding a large group of leaderless 27th Division doughboys, Sasse rallied them and led them on to their objective, where they were rejoined by their own officers.[50]

If things looked confusing to Sasse and his officers, they could only have appeared more so at Monash's Australian Corps headquarters. Although the Americans had been reporting steady progress throughout the morning, and one aviator reported seeing Americans on the objective, Monash's enthusiasm was tempered by conflicting reports from his Australian division commanders, who said they were encountering fierce opposition as they advanced behind the Americans. The problem was twofold. As the inexperienced doughboys fought their way forward, they failed to mop up pockets of German resistance that had been bypassed. Furthermore, many of the Germans sought shelter in the St. Quentin Canal tunnel, and, after the lead elements passed forward, later emerged on the battlefield from hidden passageways. It was more than the casualty-riddled, underofficered American units could cope with. Realizing the severity of the situation, Monash scrapped his original plan and ordered his forces to clear the Germans out of Bony and the northern end of the tunnel, securing the remainder of the Hindenburg Line in his sector.

Monash withdrew the badly mauled 27th and 30th Divisions from the line the night of the twenty-ninth and ordered the Australians to continue the attack the next morning. It took them two additional days to reach the corps's original objective.[51]

Casualties had been especially heavy. Although it was impossible to get an accurate count of infantry casualties because so many men had become separated from their units, either isolated behind German lines or intermingled with the Australians, it was determined that the 107th Regiment lost at least 50 percent of its assault troops—about a thousand men.[52] The 301st Tank Battalion lost three officers and seventeen enlisted men killed, fifteen officers and seventy enlisted men wounded, and seven enlisted men missing in action.[53] Of the forty tanks that began the operation, two were destroyed by British mines, and sixteen were lost to artillery fire. At the end of the day's fighting, only one vehicle, the radio tank, was still operable. All of the remaining tanks were either ditched and awaiting recovery or had fallen prey to mechanical problems.[54]

Although the heavy tankers had shown courage and aggressiveness, Sasse considered their battlefield debut a failure. He was especially critical of the tendency of tank commanders to outrun the infantrymen they were supposed to be supporting.[55] While this may have been true in A Company's sector, the responsibility for C Company's pulling away from the 107th Regiment was Sasse's, for it was he who ordered the tanks to move on to the start line after being informed that the 107th was delaying its attack by one hour. Other factors contributing to the attack's failure included a lack of counterbattery artillery fire, the failure of the infantry to mop up bypassed German positions, the unfavorable wind that blew the smoke barrage back into friendly lines, and the failure of the 106th Regiment to secure the designated start line. Nevertheless, Sasse praised his men's high morale and aggressive spirit. He singled out the performance of tank commanders who, plagued by poor visibility, frequently dismounted to lead their vehicles forward. He also commended the crews who pulled the machine guns from their shattered tanks and continued to support the infantry's advance.[56]

The 301st spent the next week working around the clock in order to get as many tanks operable as possible. By 5 October, when orders came down to prepare for an operation with the 30th Division, twenty-two tanks were ready for action.[57]

The Australians had finished clearing the Germans out of the St. Quentin Canal area and had captured the Beaurevoir Line before turning the sector over to Maj. Gen. George W. Read's American II Corps on 6 October. Read was instructed to keep up the pressure on the Germans, and this he did, using Maj. Gen. Edward M. Lewis's 30th Division.[58]

After conferring with Lewis on the evening of the fifth, Sasse moved the battalion nine miles—from its assembly area near St. Emilie to Bellicourt—to begin preparations for an attack on Brancourt scheduled for the morning of 7 October. The 301st's reconnaissance officers were briefed on the overall plan and, under the direction of Lieutenants Reynell and Naedele, moved out ahead of the rest of the battalion to begin preparing routes for the approach marches and beyond the start line.[59]

"Z" day was pushed back one day on the sixth, giving the 30th Division's staff an extra day to develop a detailed operations plan. The final plan called for the division's 59th Brigade to lead the assault with the 117th Infantry Regiment on the left (north) and the 118th Infantry Regiment on the right. Both regiments were to employ two assault battalions and retain one battalion in reserve. The 301st would divide its tanks equally between the two regiments, and the 4th Tank Brigade allotted two companies of Whippet light tanks from the British 6th Tank Battalion for exploitation with the reserve infantry battalions beyond the initial objective.[60]

Although the 30th Division had been supported by British tanks in the St. Quentin Canal attack, its lack of success left many of the infantry commanders skeptical about the value of tanks. To counteract this attitude, Sasse ordered company commanders and platoon leaders to spend as much time as possible with their infantry counterparts working out detailed coordination measures and ensuring that everyone understood the signals that would be employed to control maneuver. He would have preferred to conduct a combined training exercise, but the limited time available before the attack was to commence precluded it.[61]

All was in order by the evening of the seventh, and twenty-three tanks moved out for the lying-up point at 7 P.M. The approach march had to be made slowly as the night was very dark, and there were several wide trenches and sunken roads along the route that had to be crossed. Reconnaissance officers had partially taped the seven-mile route, relying on a parallel telegraph line to guide the unit the rest of the way. The tanks arrived at the forward position at 1 A.M. on the eighth, and crews spent the next three hours topping off fuel tanks and making final preparations for the assault, while reconnaissance officers escorted platoon leaders forward to show them the route to the start line.

Unlike the St. Quentin Canal operation, the Brancourt attack was con-

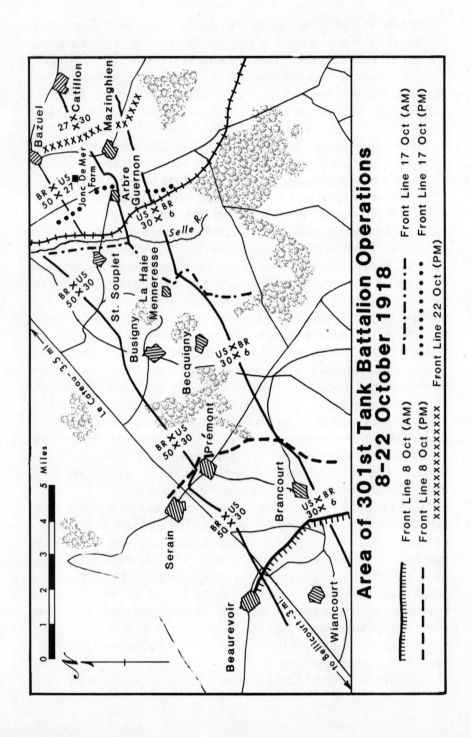

**Area of 301st Tank Battalion Operations
8-22 October 1918**

Bazuel
Catillon
27 X 30
Mazinghien
XXXXXXXX
XXXXXXX
Jonc De Mer Farm
Arbre Guernon
BR X US
50 X 27
US X BR
30 X 6
Selle R.
St. Souplet
La Haie Menneresse
BR X US
50 X 30
Busigny
Becquigny
US X BR
30 X 6
Prémont
BR X US
50 X 30
Brancourt
Serain
BR X US
50 X 30
US X BR
30 X 6
Beaurevoir
Wiancourt
to Bellicourt - 3 mi.
Le Cateau - 3.5 mi.

N

0 1 2 3 4 5 Miles

Front Line 8 Oct (AM) ———
Front Line 8 Oct (PM) ----
Front Line 8 Oct (PM) xxxxxxxxxxxxx

Front Line 17 Oct (AM) —·—·—
Front Line 17 Oct (PM) ·········
Front Line 22 Oct (PM) ———

ducted without an artillery preparation. The plan did call for a creeping barrage set to commence at zero hour, so Sasse ordered the battalion forward at 4 A.M. to ensure that the tanks covered the 2,500 yards to the start line before 5:10 A.M. Luckily, only three tanks went down with mechanical problems, and the Germans cooperated by not shelling the column and by employing a searchlight in the rear of the objective area, giving the tankers a beacon by which to guide.[62]

It was apparent from the beginning of the attack that the tankers' luck had changed. The battalion reached the start line as the friendly artillery barrage began pummeling the forward German positions. Weather conditions were perfect, with the wind carrying the artillery smoke screen east into the enemy's trenches. The tanks quickly fanned out into platoon formations, deploying ahead of the infantry with three tanks in the lead and two behind the assault element. Ground conditions were also ideal: The terrain was relatively flat and devoid of trenches or sunken roads. Two railroad lines crossed the sector, but these posed no obstacle to the tanks — although they would have been formidable ones for the infantry, as the Germans had emplaced numerous machine-gun positions along them.[63]

Second Lt. Paul S. Haimbaugh, a tank commander in Company C, displayed the exuberance many of the tankers felt in the following account of the operation:

> The doughboys spring to their feet and start forward. You urge your tank on until you are nosing the barrage; ahead the German distress signals flare in the sky. In a moment the barrage will fall. . . . Here it is and it's disconcertingly close. You think maybe you'd better zigzag a bit; maybe you can dodge 'em. The doughboys trudge sturdily on and here and there one sags into a heap (shell splinters). One shell nearly gets you as it bursts nearby with the rendering crash peculiar to high explosive. Seems like it nearly lifts your tank into the air. A dozen pneumatic hammers play a tattoo on the sides and front of your tank, and splashes of hot metal enter the cracks and sting your face and hands. The infantry, a hundred yards back, is prone on the ground; too hot for them.
>
> Well, it's up to you to locate the enemy machine guns and put them out. Observation from the peephole reveals nothing. You pop your head through the trapdoor and take a quick look around. There they are! A hasty command to the six-pounder and the machine gunners. Crash! goes the port six-pounder and the tank is filled with the fumes of cordite. A hit! A couple more in the same place and a belt of machine gun cartridges suffices to quiet that machine-gun nest.

"Come on, infantry!" As the tank passes you see the gray forms sprawled grotesquely around their guns. You are glad the bus is a male, for these six-pounders sure do the work.

Up ahead is a railroad embankment and a sunken road, a likely place for machine-gun nests. Tat-tat-tat! They've already begun to strafe you. Slipping from your seat you shout commands to your gunners. Picking their targets they pepper away with machine guns and six-pounders. The noise is terrific and the tank is filled with cordite and gasoline fumes. There is a sickening smell of hot oil about. You are pretty close now, so you order case shot and the six-pounders rake the embankment and the road with deadly effect. The place is a shambles—gray forms sprawled in the road, huddled in gun holes, lying in position about their guns. It's war, and you had to get them first. Half a dozen Germans scramble to their feet with their hands upraised and you let them pass to the rear. "Come on, infantry." Your tank surmounts the embankment and your hair raises for on a ridge 500 yards ahead are two 77s, sacrifice guns left to get you. Crash! A shell lands fifty yards to your left and short. Your starboard six-pounder lets go. Crash! Another which lands close sends a shower of stones and dirt against the sides of the tank. Working like mad, your gunner sends four shells after the two guns. Good work, for they are silenced. One member of the gun crews is able to run away. Beads of perspiration stand on your forehead. Hot work this. The combination of powder and gasoline fumes, the smell of hot oil, [and] the exhaust begins to daze you, but you pull yourself together and rumble on. The infantry swings along behind, bombing dugouts, and mopping up, assisted by your running mate, a female tank armed only with machine guns.

It's a mile to your objective now, but it's a mile of thrills. You get shot-up, put out a half a dozen machine-gun nests; clean up another sunken road with machine guns placed every ten feet along it. A one-pounder in a hedge scares you with several well-placed shots before it "goes west" [is knocked out and the crew killed]. Some German artillery observer, way back, spots you and chases you over the landscape, dropping a shell, now in front, then in back of you. Here is your objective! You wait for the infantry to come up and your crew enjoys a breath of fresh air. After the infantry is dug in and consolidated its position, you turn toward the rallying point.[64]

After clearing the railway embankments, the battalion pressed the attack on Brancourt by assaulting the village from the northern and southern flanks. One C Company crew used its six-pounder to knock out a German gun located in a church steeple, and another captured a battery of 77mm guns by attacking the position from the rear. Several tank commanders, caught up in the excitement of their success, disregarded orders to halt at Brancourt, and continued on to the second objective with the exploitation force.[65]

Cpl. Troy C. Carroll, an A Company gunner whose tank was destroyed by four direct artillery hits, ignored his own painful wounds in the back and leg and walked three miles to the rear in search of medical assistance for his crew. He then led a team of stretcher bearers back to the tank, helping them with the evacuation of the crew until he collapsed.[66]

The attack on Brancourt was a morale-boosting success for the men of the 301st. Although little more than half of the tanks reached the objective, nine of the twenty that began the attack having fallen prey to mechanical failure or German artillery fire, casualties were relatively light: two enlisted men killed and three officers and thirteen enlisted men wounded. Sasse gave high marks to the doughboys for the excellent manner in which they cooperated with his tanks, and they in turn praised the tankers. The presence on the battlefield of large numbers of German bodies riddled with case shot from the tanks' six-pounder guns also bore mute testimony to the tankers' effectiveness.[67]

The 301st withdrew to Bellicourt to rest and refit while the 30th Division continued to push the Germans back to the Selle River, then shifted its base to Serain. The 27th Division took over the II Corps front on 12 October, consolidating the position for several days while the battle-weary 30th reorganized. Orders came down on the fifteenth alerting the corps to prepare for an assault on the Selle as part of a general offensive by the British Fourth Army set to kick off on 17 October. The overall plan called for the Fourth Army to attack with three corps abreast—the British IX and XIII Corps and the American II Corps. The 4th Tank Brigade again assigned the 301st Tank Battalion to the II Corps, while the British 6th and 16th Tank Battalions would support IX Corps operations, and the British 1st Tank Battalion was assigned to the XIII Corps. The British 10th Tank Battalion was ordered into Fourth Army reserve.

On the evening of 15 October the 301st moved slightly more than four miles from Serain to Busigny with twenty-five tanks operational and began final preparations for the assault. Sasse, after consultation with the II Corps operations staff, assigned Company A to support the 27th Division in the north half of the corps sector with ten tanks. Companies B and C were consolidated under the command of Captain Clarke and ordered to support the 30th Division attack with the battalion's remaining fifteen tanks. Because the Selle River was generally within 500 yards of the jump-off line, Sasse told the infantry commanders he would hold his tanks back until they secured river crossings, then send them forward to take the lead. This directive would apply in all but the extreme southern portion of the corps sector, where the Selle ran more than a thousand yards from the start line and where numerous German strong points had been pinpointed west of the river. Here, Sasse ordered Clarke to assign one platoon to assist the infantry in their initial assault.[68]

Reconnaissance activities followed the same pattern as in previous operations, although this time the battalion's assembly area was close enough to the front to permit all of the tank commanders and platoon leaders to view the zone of attack from forward positions. In addition, all tank commanders were provided with maps marked to show river crossing sites and obstacles.[69]

The battalion left Busigny at midnight on the sixteenth, moving about five miles to a concealed position behind the Bois de Proyart. It proved to be an inauspicious start for the operation, as five tanks went down with mechanical problems, leaving only twenty to conduct the attack. Heavy German artillery fire directed at the woods added to the tankers' discomfort while they refueled their vehicles. One soldier was wounded, but none of the tanks received any damage. Shortly after 4 A.M. the battalion moved out for the start line, Company A swinging north of the woods to join the 27th Division, and Captain Clarke's composite company heading south of La Haie Menneresse to link up with the 30th. As Clarke's column approached the start line, the headquarters radio tank burst into flame in full view of the enemy, drawing fire from numerous German machine-gun positions. The tank commander immediately headed the tank into a grove of trees, concealing the vehicle before shutting it down and abandoning it.[70]

A dense fog settled over the battlefield as dawn approached, again making it impossible to see more than a few yards. At zero hour nineteen tanks crossed the jump-off line, each tank commander again feeling as though he was on his own. The fight quickly degenerated into a series of individual actions for the tankers. Several of Clarke's tanks provided valuable assistance during the initial assault on the Selle, as tank commanders looked for areas where the doughboys were pinned down and knocked out the machine-gun positions that were holding them up. Other tanks got lost in the fog and, by the time it lifted about noon, had exhausted their fuel supplies and were forced to withdraw. The river proved to be no major obstacle. All but two of the attacking tanks made it across, one getting stuck above the designated crossing point and another becoming mired in marshy ground near the bank.[71]

The heaviest fighting occurred during A Company's attack with the 27th Division, where the battalion lost its only tank to enemy fire. A German trench mortar blasted the vehicle commanded by 2d Lt. William C. Rock, setting fire to it and blowing a track off. Rock helped drag his wounded crewmen to safety, then, armed only with his pistol, charged a machine-gun position. He was killed before he could get to the Germans. Second Lt. A. B. Maury, another A Company tank commander, despite having engine problems, and with a crew that had been gassed, managed to coax his vehicle forward, capturing a German artillery battery before his motor stopped for good. Maury soon located another tank whose crew had been badly gassed and transferred his men and ammunition to the other vehicle. They continued to press the attack in advance of the doughboys and were one of only three crews to reach the final objective in the vicinity of the road between Arbre Guernon and Le Cateau.[72]

The 301st could count only seventeen tanks operable by the close of the day's operations, and it was doubtful whether any of these could mount another attack without a complete overhaul. Although the fog and mechanical problems blunted the effectiveness of the tanks, Sasse still considered the operation a success. Cooperation between tanks and infantry had been good, and his crews had shown considerable initiative in supporting the doughboys' advance. Casualties were again light: one officer and one enlisted man killed and eight enlisted men wounded.[73]

The Selle River attack proved to be the final operation for the American II Corps. Since 27 September, the corps's two divisions had suffered more than 13,000 casualties, divided about equally between them, and they had not received any replacements. In view of their depleted condition, Rawlinson ordered them back to the rear for much-needed rest and reorganization.[74]

Such was not the case with the 301st Tank Battalion, however. The Germans were reeling, and all of the 4th Tank Brigade's units were needed to support continued British efforts to keep up the pressure along the front. The 301st had withdrawn to Busigny to rest and reorganize, when the battalion was alerted to prepare for an attack on the night of 22–23 October. Losses of both personnel and vehicles had been considerable, and Sasse, like Brett in the Meuse-Argonne campaign, was forced to consolidate his remaining tanks and crews into a single provisional company commanded by Captain Clarke.[75]

The provisional company was able to muster twelve tanks for the operation, dividing them equally into three platoons. The operation plan, which was not completed until the night of the attack, did not sit well with Sasse. It required him to allocate one platoon to support two battalions of different divisions, each having separate objectives: the left flank battalion of the British 1st Division, and the right flank battalion of the British 6th Division's 18th Brigade. His remaining platoons were assigned to support the remainder of the 18th Brigade and the 6th's 71st Brigade.

The tankers devoted what was left of the twentieth and the morning and early afternoon of the twenty-first to preparing their vehicles for the move to their lying-up point near Mazinghein. The tanks were in extremely poor condition, and crews reported that many equipment items were either defective or missing.[76]

At 2:30 P.M. on 21 October, the twelve tanks of the provisional company made a three-hour road march to the lying-up point, where fuel tanks were topped off and maintenance activities continued.

Sasse, who placed considerable stock in reconnaissance and coordination, was particularly frustrated by the lack of both prior to the operation. Reconnaissance was hampered by the lack of time available, constant changes in the front-line trace, and the lack of detailed orders. The battalion's shortage of motor transport, combined with the distance between infantry brigade headquarters, limited coordination to a single briefing for platoon leaders by their respective brigade commanders on the evening of the twenty-second. To make matters worse, platoon leaders did not arrive back at the lying-up point to brief their tank commanders until two hours before the final approach march commenced, making it difficult for them to disseminate the required information.[77]

The tanks moved out for the jump-off line at midnight, reaching that position at zero hour, 1:20 A.M. on the twenty-third. The night was clear, and no artillery preparation was employed in an effort to take the Germans by surprise. Each of the platoons supporting the 71st and 18th Brigades in the left and center sectors deployed with three tanks leading the assault and one in reserve. Resistance was light, and all tanks reached their objectives in the vicinity of the towns of Bazuel, Catillons, and Ors. Sasse reported later that the creeping artillery barrage that began at zero hour was especially effective, contributing significantly to the success of the operation. Two of the tanks operating with the 71st Brigade captured eighteen Germans, turning them over to the infantry at the objective.[78]

The mission of the platoon supporting the right and left flank battalions of the 6th and 1st Divisions was confusing, and coordination with supporting artillery units was virtually nonexistent. One British artillery battery

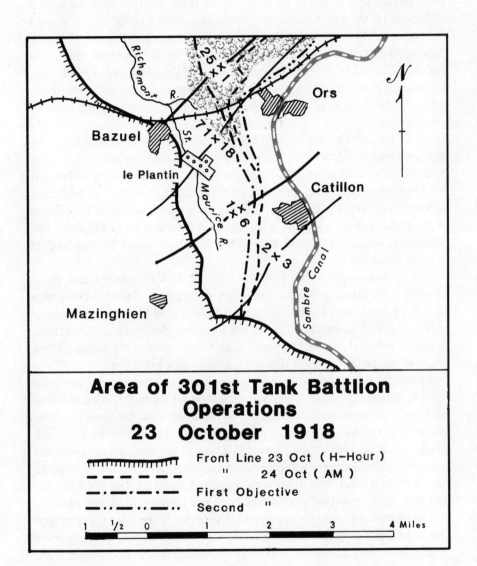

Area of 301st Tank Battlion
Operations
23 October 1918

Front Line 23 Oct (H–Hour)
"   24 Oct (AM)
First Objective
Second   "

1   ½ 0   1   2   3   4 Miles

fired only after the tank platoon leader supplied the battery commander with a barrage map containing the timetable for the creeping barrage. The tanks in this sector, deployed in the same three-up, one-back formation employed by the other platoons, crossed the start line at 1:22 A.M. and were greeted by an intense German artillery barrage. One of the tanks became lost after its compass was shattered by shell fragments, and a second tank ditched in a trench 300 yards short of the objective. The remaining assault tank reached the objective without encountering opposition, patrolling the area until ordered back to the rallying point by the infantry commander.[79]

All of the 301st's tanks rallied back at the lying-up point by 3:30 P.M. on the twenty-third, with the exception of the ditched vehicle, which had to be dug out the next day. The battalion had suffered no vehicle or personnel losses. Nevertheless, the attack of 23 October proved to be the heavy tankers' last combat action. Both vehicles and men had reached the limits of their endurance, and they were finally withdrawn from the line. The Armistice would be signed before the battalion could be restored to fighting condition.

The account of this final battle is necessarily less detailed than those of earlier operations. As the campaign progressed, Sasse's after-action reports became increasingly sketchy. It was impossible to pinpoint exact objectives, unit boundaries, and other locations described in this action, as Sasse referred to them by map coordinates only. Unfortunately, the maps to which he referred did not survive with his report.

Although results of the heavy tank operations were mixed, the 301st proved, especially in the Brancourt attack on 8 October, that tanks could be a valuable asset. Sasse repeatedly stressed the need for close cooperation between tanks and the units they supported, and his views were vindicated. The Brancourt attack showed just how formidable the tank-infantry team could be. Conversely, the battalion's poor showings in the 29 September and 23 October operations were largely a result of poor coordination and execution, especially on the part of the infantry units involved.

The heavy tanks also proved to be far superior to the light tanks in their ability to cross trenches and overcome wire obstacles and in the amount of firepower they could bring to bear on the enemy. It is ironic, in view of this obvious distinction, that virtually all of the post–World War I tank advocates would argue in favor of a doctrine to be executed by light, fast, highly mobile tanks. This is probably attributable to the fact that most of the Tank Corps's senior leaders were cavalrymen. They were frustrated by the ponderous pace of the World War I generation of tanks and hoped that light, fast armored vehicles would solve the mobility problem on future battlefields. Another limiting factor contributing to the postwar fixation on light tanks was, no doubt, the frustrating funding constraints imposed by the penny-pinching legislators of the twenties and thirties.

Finally, the tankers proved the value of personal reconnaissance. Whenever possible, leaders, including individual tank commanders, should have the opportunity to physically see the ground over which they will maneuver. Maps and photographs are valuable aids but a poor substitute for personal observation. This last lesson is one that is as important for today's armored leaders as it was for Tank Corps officers in 1918.

Notes

1. 2d Lt. John A. Logan, "The Heavy Tankers," undated, Patton Military Papers, Box 47, Personal Experience Reports of Tank Operations–1918, Patton Collection.
2. Questionnaire, Sgt. David A. Pyle, 301st Tank Battalion, World War I Survey Collection, Archives, USAMHI.
3. Coffman, *War to End All Wars,* 291–93.
4. Maj. Ralph I. Sasse, "Report on Operations: Sept. 27th–Oct. 1st, 1918," included as Appendix 8 in Rockenbach, "Tank Corps Operations," 1–2.
5. Reynell, "Personal Experience."
6. Jones et al., *The Fighting Tanks,* 31, 34.
7. Sasse, "Report on Operations," 5.
8. Sasse, "Report on Operations," 5; Reynell, "Personal Experience."
9. Sasse, "Report on Operations," 6.
10. Ibid., Appendix B.
11. Ibid., 2–3.
12. Reynell, "Personal Experience."
13. Sasse, "Report on Operations," 6.
14. Coffman, *War to End All Wars,* 293–95.
15. Sasse, "Report on Operations," 3; Reynell, "Personal Experience."
16. Reynell, "Personal Experience."
17. Sasse, "Report on Operations," 4, 6.
18. Ibid., 4; Reynell, "Personal Experience."
19. Sgt. Carl Rosenhagen, "A Day in the Life of a Tanker," *Armor* 82, no. 5 (Sept.–Oct. 1973): 34.
20. Reynell, "Personal Experience"; Sasse, "Report on Operations," 7.
21. Sasse, "Report on Operations," 6.
22. Rosenhagen, "A Day in the Life," 34.
23. Sasse, "Report on Operations," 7.
24. Rosenhagen, "A Day in the Life," 34.
25. Capt. R. deP. Clarke, "Personal Narrative," undated, Patton Military Papers, Box 47, Personal Experience Reports of Tank Operations–1918, Patton Collection.
26. General Orders no. 2, GHQ–Tank Corps, AEF, Subject: Award of Distinguished Service Cross; Reynell, "Personal Experience"; *301st Battalion History,* 173.
27. Reynell, "Personal Experience"; Sasse, "Report on Operations," Appendix B.
28. Sasse, "Report on Operations," 8.
29. Ibid., Appendix B.

30. General Orders no. 2, GHQ–Tank Corps, AEF, Subject: Award of Distinguished Service Cross.
31. *301st Battalion History,* 214.
32. Sasse, "Report on Operations," 8; Reynell, "Personal Experience."
33. Logan, "The Heavy Tankers."
34. Sasse, "Report on Operations," 8–9.
35. Logan, "The Heavy Tankers."
36. Reynell, "Personal Experience."
37. *301st Battalion History,* 198; Sasse, "Report on Operations," Appendix B.
38. General Orders no. 2, GHQ–Tank Corps, AEF, Subject: Award of Distinguished Service Cross; *301st Battalion History,* 205.
39. Ibid., 201–2; Sasse, "Report on Operations," Appendix B.
40. Clarke, "Personal Narrative"; Sasse, "Report on Operations," Appendix B.
41. Clarke, "Personal Narrative"; Logan, "The Heavy Tankers."
42. Clarke, "Personal Narrative."
43. Sasse, "Report on Operations," Appendix B.
44. Rosenhagen, "A Day in the Life," 35.
45. Ibid.
46. Ibid.
47. Clarke, "Personal Narrative."
48. Sasse, "Report on Operations," Appendix B.
49. *301st Battalion History,* 198–201.
50. Ibid., 177; Reynell, "Personal Experience."
51. Coffman, *War to End All Wars,* 296–97.
52. Ibid., 297.
53. Sasse, "Report on Operations," Appendix A.
54. Ibid., Appendix B.
55. Ibid., 10.
56. Ibid.
57. Maj. Ralph I. Sasse, "Report on Operations October 8, 1918," reprinted in *301st Battalion History,* 97.
58. Coffman, *War to End All Wars,* 297.
59. Sasse, "Operations, October 8," 96; Reynell, "Personal Experience."
60. Sasse, "Operations, October 8," 96.
61. Ibid.
62. Ibid., 98.
63. Ibid., 102.
64. 2d Lt. Paul S. Haimbaugh, as quoted in Arch Whitehouse, *Tank: The Story of Their Battles and the Men Who Drove Them from Their First Use in WWI to Korea* (Garden City, N.Y.: Doubleday & Co., Inc., 1960), 115–16.
65. Sasse, "Operations, October 8," 102; Clarke, "Personal Narrative."

66. General Orders no. 2, GHQ–Tank Corps, AEF, Subject: Award of Distinguished Service Cross.
67. Sasse, "Operations, October 8," 102.
68. Maj. Ralph I. Sasse, "Report on Operations, October 17, 1918," reprinted in *301st Battalion History*, 112–14.
69. Ibid., 112–14.
70. Ibid., 114; Clarke, "Personal Narrative."
71. Sasse, "Operations, October 17," 114–18.
72. Ibid., General Orders no. 2, GHQ–Tank Corps, AEF, Subject: Award of Distinguished Service Cross.
73. Sasse, "Operations, October 17," 118.
74. Coffman, *War to End All Wars*, 297–98.
75. Sasse, "Report of Action of October 23, 1918," reprinted in *301st Battalion History*, 132; Clarke, "Personal Narrative."
76. Sasse, "Action, October 23," 132–34.
77. Ibid., 134.
78. Ibid., 136.
79. Ibid.

Epilogue

The personnel strength of the Tank Corps had reached its peak by the time the Armistice was signed, the AEF Tank Corps counting 752 officers and 11,277 enlisted men in its ranks, while an additional 483 officers and 7,700 enlisted men manned Tank Corps units in the United States—divided about equally between Camps Colt and Polk.[1]

Although the Tank Corps was authorized a total of 1,803 officers and 25,535 enlisted men on 11 November 1918, the War Department took quick action to reduce the ceiling to 300 officers and 5,000 enlisted men. These were to be retained in three brigades, a tank center with a repair and salvage company and depot company, and the Tank Corps's general headquarters.[2]

Rapid demobilization soon followed. Looking back on the experience, Eisenhower said, "Nothing at West Point or in the forty months since graduation had prepared me for helping to collapse an Army from millions to a peacetime core. The new problem kept us even busier than we had been in the middle of summer."[3]

The greatest challenge facing the young Camp Colt commander was maintaining discipline and morale. Eisenhower stressed to his subordinate leaders that they would have to exercise considerable restraint in their dealings with the citizen soldiers anxiously awaiting discharge. The key, in his view, was to develop "the ability to explain, reasonably and persuasively, all the necessary measures that would have to be taken before we could fold up the camp and the war." Eisenhower saw make-work programs as "self-defeating" and the continuance of intensive training as "ridiculous." Nevertheless, as the training pace slowed, housekeeping activities intensified. The troops were mollified, however, by the knowledge that they would be moving to Camp Dix, New Jersey, in a few weeks for discharge.[4]

The Camp Colt contingent arrived at Dix in early December, and Eisenhower received word that he was to discharge all but 250 enlisted men. Colonel Clopton, the Tank Corps commander at Camp Polk, was instructed to retain 200 men. Only thirty-one officers from Tank Corps units in the United States were to be retained. As soon as these strengths were met, Camp Polk was to be abandoned and the remaining tankers moved from there and Camp Dix to Camp Benning, Georgia.[5]

The one-page "history" of the 343d Light Tank Battalion bears mute testimony to the rapidity of the Tank Corps's buildup and demobilization. This document, signed by the battalion commander, 1st Lt. G. A. Bauman, shows that the unit existed for less than two months. Organized on 9 November 1918 at Camp Polk, the battalion transferred to Camp Greene, near Charlotte, North Carolina, on 5 December, and was demobilized on 29 December.[6]

The AEF Tank Corps labored in a similar atmosphere. Although a few tank battalions were ordered to perform occupation duty, the majority gathered at the 302d Tank Center at Bourg to prepare for their return to the United States and demobilization. During the waning days of 1918 and the early part of 1919, Patton found peacetime activities to be "almost the same as usual" but added that "it is not so easy to get up in the morning as it used to be especially on saturday. I fear that laziness which ever pursued me is closing in on me at last."[7]

While his tankers continued to train and perform demonstrations for infantry units and the AEF schools, Patton busied himself with writing after-action reports and a history of the 304th Tank Brigade and with giving lectures.

Patton enumerated nine "Tactical Conclusions" drawn from the 304th Brigade's participation in the Meuse-Argonne campaign in his after-action report—conclusions that were similar to those expressed by both Brett and Sasse in their reports. Heading the list was a critical comment directed against senior officers who "in their demands on the tanks did not seem to realize their limitations and especially the fact that tanks must have infantry operating with them, if they are to be successfully employed."[8] What makes this observation particularly insightful is the subtle implication—probably unconscious, as Patton biographer Martin Blumenson notes—that tanks were less an infantry support weapon than an offensive system requiring infantry support to exploit its capabilities. Despite this implication, Patton continued to adhere to the "party line" in the ensuing months, thus inhibiting his attempts to achieve a radical breakthrough in tank doctrine.[9]

Patton further concluded that:

- A lack of liaison between tanks and infantry severely handicapped the tanks during operations.
- The infantry used the tanks as a crutch, expecting them to overcome enemy resistance and consolidate objectives after successful attacks.
- Tanks, because of their mechanical weaknesses, should not be squandered in a reconnaissance role.
- The distance between attack positions and lines of departure should be reduced in order to cut losses due to mechanical failure.
- There is no substitute for physical ground reconnaissance by key leaders.
- Measures such as smoke screens and dedicated artillery units for counterbattery fire should be employed to reduce the effectiveness of enemy artillery against tanks.
- Tanks clearly demonstrated their value as offensive weapons and as a separate combat arm.
- Changes in tactics, especially with regard to better use of tanks in mass and depth, were needed.[10]

In March 1919, Tank Corps units from overseas began arriving at Camp Meade, Maryland, where they were joined by Eisenhower and the remaining stateside tankers who had gathered at Camp Benning. Meade was to be the "home" of the Tank Corps and soon housed the Corps's entire complement in the Benjamin Franklin Cantonment. The force consisted of the 304th Tank Brigade with the 303d Heavy Tank Battalion, 328th and 344th Light Tank Battalions, and 321st Repair and Salvage Company; the 305th Tank Brigade with the 301st Heavy Tank Battalion, 329th and 330th Light Tank Battalions, and 306th Repair and Salvage Company; the 306th Tank Brigade with the 306th Heavy Tank Battalion, 331st and 345th Light Tank Battalions, and 316th Repair and Salvage Company; and the 302d Tank Center with the 317th Repair and Salvage Company and a provisional depot company.[11]

Because Brigadier General Rockenbach remained in France to supervise the redeployment of the remnants of the AEF Tank Corps, Colonel Welborn retained his position in Washington as the stateside Tank Corps director. Colonel Clopton moved to Meade as post commander and also served as Welborn's deputy.[12]

Tank production efforts, which had failed miserably during the war, improved considerably in 1919. Components for the Mark VIII Liberty heavy tank were shipped to Rock Island Arsenal, and a total of 100 were built by March. The Maxwell Motor Company and affiliated contractors, who had completed sixty-four copies of the Renault light tank by the time the Armistice was signed, built an additional 714 by 31 March, and total production reached 952 before contracts were canceled later in the year. In addition, Tank Corps units redeploying from France brought back 218 French-built Renaults and twenty-eight British Mark Vs.[13]

While training operations and demonstrations kept the rank-and-file tankers occupied at Meade during 1919 and 1920, Tank Corps officers entered into a protracted campaign to ensure the Corps's survival as a separate combat arm.

In mid-April 1919, Patton, Lt. Cols. James E. Ware and Joseph Viner, and Maj. Ralph Sasse were appointed members of a newly created Tank Board charged with formulating tank regulations, writing a drill manual for tank crews, and preparing a course of instruction for tankers. After spending a month in Washington working out the basic doctrine for the Tank Corps, the members of the Tank Board were joined by Col. Henry Mitchell for an inspection tour of tank production and research facilities at Rock Island Arsenal, Illinois, and the Springfield Armory in Massachusetts.[14]

Congress, meanwhile, was engaged in a debate that would have significant impact on the future of the Army and the Tank Corps. The backlash against war and all things military was being felt all across America, and Congress, sensing the public mood, became increasingly penurious. Word quickly filtered through the ranks that a major reduction in force structure was forthcoming and that the majority of Class III officers—those who had been commissioned from the enlisted ranks during the war—would be eliminated and regular officers restored to their permanent prewar ranks. The cuts ordered for the Tank Corps were particularly distressing. On 11 July 1919, Congress passed a law that included a reduction in the Corps's strength to 154 officers and 2,508 enlisted men.[15]

In August, with the AEF Tank Corps's redeployment all but complete,

Secretary of War Newton D. Baker ordered Rockenbach to report to Camp Meade and take over the Tank Corps. As commandant, Rockenbach was in a particularly good position to express his views. Unfortunately for the Tank Corps, the methodical, hidebound general was not the right man for the job. He was more interested in maintaining the status quo than in promoting research, development, and training—three essentials for the creation of a vigorous, improving force.[16]

Careful examination of Rockenbach's postwar lectures to the General Staff and officers at the Army War College bears out his subordinates' charge that he was narrow-minded and lacked originality. None of the Rockenbach lectures contained anything new. They were, in fact, largely verbatim transcripts of his AEF operations report drafted in December 1918. This proved particularly frustrating to tank officers like Patton and Eisenhower, who were becoming more and more convinced that the Army should employ fast tanks similar to those already in production in England and France in mass formations to achieve breakthroughs and lead in the exploitation.

Maj. Gen. Charles P. Summerall, whose 1st Infantry Division and V Corps had been beneficiaries of tank support in the Meuse-Argonne campaign, was especially critical of the Tank Corps's chief. Following Rockenbach's 1919 lecture to the General Staff, Summerall wrote, "Far from disagreeing with any part of the lecture, the only comment that I heard . . . was that you had presented the subject in a very conservative manner, and that all were in hearty sympathy with the development and use of the Tank Corps."[17]

But Summerall and Chief of Staff Gen. Peyton C. March were among the few senior officers who were convinced of the offensive value of tanks and who favored retention of the Tank Corps as a separate arm. General Pershing was the leading proponent of a plan to place the tankers under the control of the Infantry Branch, and his view was bolstered by the findings of a War Department board convened in 1919 to study tank tactics. While the board members recognized the value of tanks in support of the infantry, they declared tanks to be incapable of independent operations. The board concluded that the "Tank Service should be under the general supervision of the Chief of Infantry and should not constitute an independent service."[18]

But the tankers kept the debate lively. Eisenhower and Major Brett, who had spent the summer as Tank Corps observers on the Army's First Transcontinental Motor Convoy of 1919, a 3,251-mile trek from Washington to

San Francisco that lasted from 7 July to 5 September, returned to Meade in the fall, where they joined in the discourse on future tank tactics and doctrine fostered by Patton and a group of young tank officers who were discouraged by Rockenbach's efforts to maintain the status quo, and to whom affiliation with the infantry was anathema.

According to Eisenhower, the members of this small clique thought:

> Tanks could have a more valuable and more spectacular role. We believed that they could be speedy, that they should attack by surprise and in mass. By making good use of terrain in advance, they could break into the enemy's defensive positions, cause confusion, and by taking the enemy front line in reverse, make possible not only an advance by infantry, but envelopments of, or actual breakthroughs in, whole defensive positions.[19]

By early 1920, Patton, Eisenhower, and others had refined their tactical ideas and described in great detail the characteristics of the tank they thought would be best capable of exploiting such doctrine. The tank they wanted had to be fast and reliable and to pack considerable firepower. Armor should be sufficient to protect the crew and critical parts from machine-gun and light-cannon fire but not so heavy as to sacrifice mobility.[20]

They found additional help in the tactical problems used in the Command and General Staff School, which Patton had obtained to help him prepare for the Leavenworth course. The officers would first solve the problems on their own, then compare their answers to the "school solutions" approved by the Leavenworth faculty. Once this had been accomplished they would tackle the problems again, this time "adding to the forces involved a complement of our dream tanks. . . . The troops supported by tanks always 'won' in our revised solution," Eisenhower recalled.[21]

Both Eisenhower and Patton wrote articles expounding their theories for the *Infantry Journal* in mid-1920. But before their papers reached print, Rockenbach fired the Tank Corps's opening salvo at the forces hoping to make it an adjunct of the infantry. The Rockenbach article appeared in two parts in the January and February editions of the *Infantry Journal*. It was little more than a rehash of the tank-infantry tactics worked out during the world war and included a conservative plea that the Tank Corps remain a separate arm so it could better support all branches rather than be drawn into an entangling alliance with any single branch.[22]

Patton's article, which appeared in the May issue under the title "Tanks in Future Wars," was highly provocative. From the beginning, it appeared he was determined to alienate his audience, taking aim, as he did, at "inexperienced officers [who] 'poo-poo' the idea that tanks will ever again be used, or that, if used, they will have any material effect."[23]

According to Patton, the infantry's lack of interest in tanks could be traced to the relatively few American units that operated with tanks, the tanks' own mechanical deficiencies, and the fact that trench warfare was particularly ill suited for tank operations. He noted, however, that the tanks' greatest successes had been achieved after July 1918, when "trench warfare was practically extinct," and that the second generation of tanks coming off overseas production lines was vastly superior to "the grotesque war babies of 1918."[24]

The heart of Patton's article, however, is his impassioned response to "unimaginative conceptions" of future conflict:

> Too many people vainly fancy that future unpleasantness will follow as sealed patterns of the World War—with trenches, barrages and plasticine maps and with air photographs so accurate that the latest activities of some careless rabbit are easily discernible. Wars of preparations, concentrations and chocolates; of air raids, welfare workers and "Big Berthas." Above all, of wars with endless trenchments and flankless armies. They forget that not in Asia, Africa or America can such a war be staged. Because all the above luxuries depend upon two things: First, roads—hundreds of good metalled roads to carry the limitless supplies; second, fronts short enough to be continuously occupied, and "get-at-able" enough to allow for the feeding of the garrison.
>
> In the continents just mentioned the scanty network of inferior roads precludes the first, and their vast size prevents the second of the above essentials.[25]

Patton dismissed the notion of future conflict in Europe with the observation that "the improved tank, using tactics already proven, is of undoubted tactical value" and focused his attention on possible operations "in the vast continents of the 'A's.'" Combat in such an environment would be highly mobile, Patton wrote. Artillery would be spread thinly, capable of little more than counterbattery fire or engaging targets of opportunity. The infantry would be forced to rely on its own fire superiority to advance, and if the attacking force "has the élan and the number; the defender, countless machine guns, hidden on reverse slopes and undisturbed by the barrage of former days; the struggle will be long and very slow."[26] In Patton's mind, the tank offered the solution. He demonstrated how tanks might be employed in rear or flank guard roles, in the defense, and, most importantly, as offensive weapons.

Patton tempered his conclusion with the caveat that "there is no belief on the part of any tank officer that the tank has replaced in the least degree any one of the existing arms." Nevertheless, he wanted readers to know that he was firm in the conviction that tanks must be maintained as a separate combat arm:

> The tank is new and, for the fulfillment of its destiny, it must remain independent. Not desiring or attempting to supplant infantry, cavalry, or artillery, it has no appetite to be absorbed by any of them.
>
> Our A.E.F. colors—Red, Yellow and Blue—were happily chosen and are truly significant. We have the cannon of the artillery—the machine gun of the infantry, and the crushing power and mobility of the cavalry horse. As an independent corps, we may assist any one of the arms as directed. Absorbed by any one of them, we become the stepchild of that arm and the incompetent assistant of either of the others. . . .
>
> Like the air service, [tanks] are destined for a separate existence.
>
> The tank corps grafted on infantry, cavalry, artillery, or engineers, will be like the third leg to a duck—worthless for control, for combat impotent.[27]

Shortly after Patton's article appeared, the debate over the Tank Corps's future reached its climax. While the Army's top leadership was split on the issue, it was a simple question of economics for Congress. Congressman Harry E. Hull of Iowa summed up the congressional viewpoint when he observed that "perhaps in the case of war there might be some need of a separate organization for tanks, but I am unable absolutely to see any reason during peacetime for the creation of the overhead that would have to be established to give you a separate organization."[28]

The axe fell on 2 June 1920, when Congress passed the National Defense Act, Section 17 of which dissolved the Tank Corps as a separate entity, assigned all tank units to the infantry, and directed that tankers be designated as serving in the "Infantry (Tanks)." It was a harsh blow to the officers and men of the Tank Corps, but they were hardly alone. The legislation contained numerous provisions impacting negatively on large segments of the Army. Hardest hit was the Officer Corps. Regulars would revert to their prewar ranks at the end of June, and boards were established to identify all but a handful of temporary officers for separation. The legislation also called for postwar reorganization of the Army extending far beyond the elimination of the Tank Corps. Proponents of air power, led by the flamboyant Brig. Gen. William "Billy" Mitchell, succeeded where the tankers failed, convincing Congress to establish a separate Air Service. The Chemical Warfare Service and Finance Department also were designated as separate branches.

Morale in the tank units plummeted. Rockenbach lost his star, reverting to the rank of colonel, and took up duties as Camp Meade commander and titular head of a branch of service that had lost its independent status. Patton and Mitchell, commanders of the 304th and 305th Tank Brigades at Meade, both lost their eagles and pinned on captains' bars at the end of the month. Mitchell, however, was promoted to major on 1 July. Out of respect for Patton's service, Rockenbach retained him in command of the 304th, assigning majors to the tank headquarters staff or to Mitchell's 305th Brigade, leaving only company-grade officers in Patton's command.[29]

By mid-August, Patton had become convinced that he had no future serving with tanks. On the fifteenth, he submitted a formal request for transfer back to the cavalry. Martin Blumenson speculates that Patton's motives were purely personal. With the Tank Corps's subordination to the Infantry Branch, Patton lost his status as a high-ranking and experienced tank officer with reasonable hopes for attaining flag rank. Most of his prominent friends, from Pershing down, were cavalrymen—few wore the crossed rifles of the infantry. Thus, his best hope for promotion lay in the pursuit of cavalry assignments. More importantly, Congress had clearly demonstrated its intent to preserve the status quo. Funding for tank research and development was virtually nil, and it appeared that, at least in the United States, tankers would be doomed to operate well into the foreseeable future with the slow, fragile, "grotesque war babies" spawned in France and England. Any hopes Patton might have harbored that tanks would be transferred to the cavalry rather than the infantry were dashed by Congress. Ironically, Patton himself helped influence that decision by continuing to tout the tank as an infantry support weapon, rather than expanding upon and more forcibly arguing his belief that new tactics and doctrine needed to be developed calling for fast tanks supported by infantry.[30]

Patton's request was approved, and on 30 September 1920 his three-year association with tanks came to an end. During the next two decades he would continue to study the progress of mechanization. In the late 1920s and early 1930s, Patton enhanced his stature in the cavalry by ardently defending tactical use of the horse, while also advocating machines – a fine line, but one he trod well.

By 1940, his career had come full circle. During the spring, Patton served as an umpire for the Third Army maneuvers, conducted in two phases at Forts Benning, Georgia, and Beauregard, Louisiana. Here he saw an ad hoc armored division composed of Brig. Gen. Adna R. Chaffee's 7th Mechanized Cavalry Brigade from Fort Knox, and a provisional motorized tank brigade from Benning, run roughshod over Maj. Gen. Kenyon A. Joyce's 1st Cavalry Division. The impact of the maneuvers on the future of the horse cavalry was not lost on Patton, who finally renounced his allegiance to that branch and actively sought command of a unit in the soon-to-be-created Armored Force. He got his wish on 27 July, when he reported to Fort Benning to take over the newly organized 2d Armored Division's 2d Armored Brigade. It was the first rung on the ladder of armored command Patton would climb to fame in the Second World War.[31]

Eisenhower, meanwhile, remained in command of a heavy tank battalion at Meade, and in November 1920 his article on tanks appeared in the *Infantry Journal*. His arguments followed along lines similar to Patton's, although characteristically couched in more tactful terms. Missing, however, is the farsighted description of mobile armored tactics. Instead, Eisenhower suggested that the infantry consider replacing its divisional motorized machine-gun battalions with a single company of tanks. This would, he argued, give the infantry division commander a hard-hitting, fast-moving force far superior to the existing machine-gun unit.[32] It was a message the infantrymen did not want to hear. Eisenhower was called on the carpet by the chief of infantry, who told him "that my ideas were not only wrong but dangerous and that henceforth I would keep them to myself. Particularly, I was not to publish anything incompatible with solid infantry doctrine. If I did, I would be hauled before a court-martial."[33]

Not long afterward, Eisenhower, too, requested transfer from tanks. Unfortunately, his plea for reassignment fell on deaf ears. Rockenbach insisted that he needed Eisenhower and sent the request to the War Department with an endorsement asking that it be denied. It was.[34]

Eisenhower was rescued from his career dilemma by Brig. Gen. Fox Conner in early 1922. Shortly after returning from France in late 1919, Conner visited Meade at Patton's invitation. Following an early dinner,

Patton and Eisenhower conducted Conner on a tour of the tank center's shops and motor park. Later, Conner fired a barrage of questions at Eisenhower, some of which could be answered easily, "while others required long explanations."[35] Conner was sufficiently impressed with Eisenhower that he went to Pershing, who became chief of staff in 1921, and asked that Eisenhower be assigned to the staff of Conner's infantry brigade in Panama. As Eisenhower later remembered, "the red tape was torn to pieces," and he got the assignment in January 1922.[36]

The next three years were especially valuable for Eisenhower. Conner has long been recognized as one of the Army's greatest intellectuals, and Eisenhower described the tutoring he received as "a sort of graduate school in military affairs and the humanities, leavened by the comments and discourses of a man who was experienced in his knowledge of men and their conduct."[37] Conner repeatedly impressed upon Eisenhower his belief that another major war was inevitable. He encouraged Eisenhower to prepare himself for that conflict and told him that one of the best ways he could do so would be to seek an assignment with Col. George C. Marshall. Conner thought Marshall was a genius and saw in Eisenhower many similar qualities. Although Eisenhower never did serve with Marshall, the two met twice, if only briefly, before World War II. More importantly, Marshall stopped in Panama en route to China in 1924, where he met with his old friend Conner. There is little doubt that Eisenhower was one of the topics of their discussion. There is equally little doubt that Marshall was sufficiently impressed by his mentor's praise of the young officer that he would not fail to remember Eisenhower's name in more troubled times.[38]

The exodus of experienced tank officers hoping to salvage their careers elsewhere continued throughout 1920 and 1921.

Joseph Viner, who reached lieutenant colonel during the war and who took over the tank center at Bourg just prior to the St. Mihiel offensive, resigned his commission in January 1921. He went on to become a business executive with the International Printing Ink Corporation of New York.[39]

Daniel D. Pullen, whose 3d Tank Brigade headquarters acted as the liaison for the French tank units at St. Mihiel and in the Meuse-Argonne campaign and who commanded the 306th Tank Brigade at the close of hostilities, returned to the Corps of Engineers. He died at Walter Reed Army Hospital in Washington on 22 September 1923 of an illness he picked up while serving as the department engineer in Panama. He was thirty-eight.[40]

Henry E. Mitchell, who commanded the initial contingent of heavy tanks in England and who went on to become a colonel and commander of first the 2d and then the 305th Tank Brigades, left tanks to serve as the assistant professor of military science and tactics at Norwich University from August to November 1920, then at the University of Kentucky until June 1921. He was the V Corps area finance officer for three years before returning to cavalry assignments during the remainder of the 1920s and early 1930s. He was promoted to colonel again on 1 May 1934 and placed on the Disability Retired List. He died on 1 August 1937 at the age of forty-nine.[41]

Ira C. Welborn, the Tank Corps director in the United States during the war, returned to the infantry. He spent two years, from June 1921 to September 1923, on the General Staff and commanded the 4th and 35th Infantry Regiments and the 22d Infantry Brigade before retiring in 1932 with more than thirty-four years of service.[42]

Ralph I. Sasse, who commanded the 301st Heavy Tank Battalion in battle, left Meade as a cavalry captain in June 1920. He spent the next two decades in various cavalry assignments and coached the West Point football team in 1928, and again from 1930 to 1933, before retiring as a lieutenant colonel in November 1940. Recalled to active duty on 8 July 1942, he was promoted to colonel on 27 October 1942 and commanded an armored group at Fort Knox in 1942–43, and the 7th Armored Regiment at Fort Meade and Camp Adair, Oregon, in 1945.[43]

The vast majority of the World War I Tank Corps's officers were temporary officers—commissioned directly from the enlisted ranks or the product of officer training schools—who became victims of the reduction in force in 1919 and 1920. This was the fate that befell men like Capts. Elgin Braine, Ranulf Compton, Harry Semmes, Courtney Barnard, William Williams, and a host of others. Nevertheless, according to Semmes, many of them "held [Patton] in friendly esteem," maintaining contact with their former commander after the war and volunteering their services to him when the Second World War began. Patton helped several return to uniform. Among these were Semmes, who commanded the 3d Armored Landing Team as a lieutenant colonel during the North Africa invasion; Arthur Snyder, who commanded a tank battalion in Italy; Horatio Rogers, who had received a battlefield commission in 1918; Loyall F. Sewall, who served in Patton's Third Army; and a World War I tank driver, Bathurst Chambliss, who fought in tanks again during the Second World War.[44]

Of the World War I Tank Corps's major figures, only Sereno E. Brett opted to maintain his allegiance to tanks during the troubled 1920s and 1930s. Brett commanded the Expeditionary Tank Force in Panama in 1923–24 after graduating from the Infantry School. In 1927 he was graduated from the Command and General Staff School at Leavenworth, then returned to Benning to serve as an instructor at the Infantry School. In 1930–31 he was executive officer of the Experimental Mechanized Force at Meade, then a member of the Infantry Board from 1931 to 1938, with a break in 1934 to attend the Army War College and to serve a brief stint with the Hawaiian Division. Brett joined the newly created Armored Force at Fort Knox as chief of staff in 1940 and commanded the 31st Armored Regiment in 1941. He joined the staff of the 5th Armored Division in 1942–43 but was forced to retire as a colonel because of physical disability before that unit deployed overseas. He died on 9 September 1952 at the age of sixty-one.[45]

Colonel Rockenbach spent the last few years of his career as commander of the few remaining tanks at Camp Meade. He continued to make feeble pleas—which fell on deaf ears—for an independent tank corps until he faded into retirement.

Thanks to the budget-cutting axe wielded by Congress, by late 1920 only two tank battalions—the 16th Light and 17th Heavy—and a maintenance company were left at Meade. At the suggestion of Maj. Gen. Charles P. Summerall, who hoped to foster closer cooperation between tanks and infantry, the secretary of war carved up the rest of the tank force, assigning one tank company to each infantry division and a battalion to the Infantry School at Fort Benning. Meade continued to house the Tank School, however, and remained the hub of postwar tank activities until that locus shifted to Fort Knox, Kentucky, in the mid-1930s.[46]

While the United States virtually ignored the tank during the remainder of the 1920s and early 1930s, progressive-thinking officers like Maj. Gen. John F. C. Fuller and Capt. Basil Liddell-Hart in England, Gen. Heinz Guderian in Germany, and Col. Charles de Gaulle in France developed bold new tactics and strategic concepts for the employment of armored vehicles. Their writings served as an inspiration to a new generation of American officers, including Daniel Van Voorhis, Adna R. Chaffee, Robert W. Grow, and a handful of others who built upon the lessons learned by the Tank Corps in World War I and who nursed the Army's comparatively small mechanized force through the long, lean years before Germany unleashed its *blitzkrieg* on a shocked world.

Notes

1. Brig. Gen. Samuel D. Rockenbach, "Report of Chief of the Tank Corps," 13 Oct. 1919, USAMHI, 4375.
2. Chief of Tank Corps to Historical Branch, General Staff, Subject: Successive Increases in the Tank Corps, Apr. 6, 1917–Nov. 11, 1918, 15 Dec. 1919, Record Group 165, Entry 310, Box 216, File 7-61.13/6, NA.
3. Eisenhower, *At Ease,* 151–52.
4. Ibid., 152.
5. The adjutant general to director, Tank Corps, Subject: Discharge, Enlisted Men, Tank Corps, 26 Nov. 1918, Record Group 165, Entry 310, Box 216, File 7-61.13/6, NA.
6. 1st Lt. G. A. Bauman, "History of the 343rd Battalion, Tank Corps," 29 Dec. 1918, Record Group 165, Entry 310, Box 447, File 66-11.4, NA.
7. George S. Patton, Jr., to Beatrice Ayer Patton, 16 Nov. 1918, as quoted in Blumenson, *Patton Papers,* 641.
8. Col. George S. Patton, Jr., "Operations of the 304th Brigade, Tank Corps from September 26th to October 15th 1918," 18 Nov. 1918, Patton Chronological Files, Box 11, 7–20 Nov. 1918, Patton Collection, 9.
9. Blumenson, *Patton Papers,* 642, 650.
10. Patton, "Operations of the 304th Brigade," 9–10.
11. Eisenhower, *At Ease,* 156; Blumenson, *Patton Papers,* 699; letter, chief of Tank Corps to Historical Branch, General Staff.
12. Blumenson, *Patton Papers,* 699.
13. Col. Leonard P. Ayers, *The War with Germany: A Statistical Summary* (Washington, D.C.: U.S. Government Printing Office, 1919), 80; Jones et al., *The Fighting Tanks,* 157; Timothy K. Nenninger, "The Tank Corps Reorganized," *Armor* 78, no. 2 (Mar.–Apr. 1969): 34.
14. Blumenson, *Patton Papers,* 705–6.
15. Chief of Tank Corps to Historical Branch, General Staff.
16. Nenninger, "The Tank Corps Reorganized," 34–35.
17. Ibid., 35.
18. Ibid.
19. Eisenhower, *At Ease,* 170.
20. Ibid.
21. Ibid., 173.
22. Brig. Gen. Samuel D. Rockenbach, "Cooperation of Tanks with Other Arms," *Infantry Journal* 16, no. 7 (Jan. 1920): 533–45; 16, no. 8 (Feb. 1920): 622–73.
23. Col. George S. Patton, Jr., "Tanks in Future Wars," *Infantry Journal* 16, no. 11 (May 1920): 958.

24. Ibid., 958–59.
25. Ibid., 959.
26. Ibid.
27. Ibid., 961–62.
28. As quoted in Nenninger, "The Tank Corps Reorganized," 36.
29. Blumenson, *Patton Papers,* 739.
30. Ibid., 741–42.
31. Ibid., 949–56.
32. Capt. Dwight D. Eisenhower, "A Tank Discussion," *Infantry Journal* 17, no. 5 (Nov. 1920): 453–58.
33. Eisenhower, *At Ease,* 174.
34. Ibid., 179.
35. Ibid., 178.
36. Ibid., 182.
37. Ibid., 187.
38. Ibid., 195; Miller, *Ike the Soldier,* 212–13.
39. *General Cullum's Biographical Register of the Officers and Graduates of the United States Military Academy,* Vol. 7, *Supplement 1920–1930* (Saginaw, Mich.: Seeman & Peters Printers, 1920), 983; ibid., Vol. 8, *Supplement 1930–1940* (Chicago: R. Donnelly & Sons Co., The Lakeside Press, 1940), 265.
40. Ibid., Vol. 7, 866.
41. Ibid., Vol. 6A, *Supplement 1910–1920,* 1000–1001; ibid., Vol. 7, 568–69; ibid., Vol. 8, 139–40.
42. Ibid., Vol. 5, *Supplement 1900–1910,* 607; ibid., Vol. 7, 470; ibid., Vol. 8, 116.
43. Ibid., Vol. 7, 1143; ibid., Vol. 8, 313.
44. Semmes, *Portrait of Patton,* 55–56.
45. *Who Was Who in American History—The Military* (Chicago: Marquis Who's Who, Inc., 1975), 64.
46. Nenninger, "The Tank Corps Reorganized," 36.

APPENDIX A

Tank Specifications [1]

French Renault Light Tank

Dimensions: Length, 16 feet, 5 inches; width, 5 feet, 8 inches; height, 7 feet, 6½ inches
Armor plate: 0.3 to 0.6 inches thick
Weight: 7.4 tons
Engine: Renault 4-cylinder, 39-horsepower, thermo-siphon cooled
Transmission: Sliding gear with 4 forward speeds and 1 reverse
Horsepower/ton ratio: 5.3
Tracks: Flat metal plates 13 inches wide, with single grousers
Suspension: Coil and leaf spring with pivoted bogies
Fuel capacity: 24 gallons
Range: 24 miles
Maximum speed: 6 miles per hour
Armament: 1 37mm gun or 1 8mm machine gun mounted in a fully traversing turret
Obstacle ability: The Renault could cross trenches 6½ feet wide, ford a stream 27 inches deep, climb a 45-degree slope, scale a 24-inch wall, and knock down trees up to 10 inches in diameter.
Crew: 2

United States Six-Ton M-1917

Dimensions: Length, 16 feet, 5 inches; width, 5 feet, 10½ inches; height, 7 feet, 7 inches
Armor plate: 0.25 to 0.6 inches thick
Weight: 7.25 tons combat loaded; 6.7 tons empty
Engine: Buda 4-cylinder, 42-horsepower, forced-water cooled
Transmission: Sliding gear with 4 forward speeds and 1 reverse

---

[1] Source: Jones et al., *The Fighting Tanks.*

Horsepower/ton ratio: 5.8

Tracks: Flat steel plates 13⅜ inches wide, with single grousers

Suspension: Coil and leaf spring with bogies and return rollers. The track frame was pivoted at the rear and sprung in front.

Fuel capacity: 30 gallons

Range: 30 miles

Maximum speed: 5.5 miles per hour

Armament: 1 37mm gun or 1 .30-caliber machine gun mounted in a fully traversing turret

Obstacle ability: The M-1917 could cross 7-foot-wide trenches, ford streams 2 feet deep, climb a 35-degree slope, scale a 3-foot wall, and knock down trees 8 inches in diameter.

Crew: 2

British Mark V

Dimensions: Length, 26 feet, 5 inches; width, (male) 13 feet, 6 inches, (female) 10 feet, 6 inches; height, 8 feet, 8 inches

Armor plate: 0.2 to 0.47 inches thick

Weight: Male, 31.9 tons; female, 31 tons

Engine: Ricardo 6-cylinder, 150-horsepower, forced-water cooled

Transmission: Planetary with 4 forward and 4 reverse speeds

Horsepower/ton ratio: Male, 4.7; female, 4.8

Tracks: Flat steel plates 26½ inches wide, with single grousers

Suspension: Rigid with rollers

Fuel capacity: 108 gallons

Range: 25 miles

Maximum speed: 4.6 miles per hour

Armament: Male, 2 6-pounder (57mm) guns, and 4 8mm machine guns; female, 6 8mm machine guns. In some female models a 6-pounder gun was substituted for 1 of the machine guns. These were called composites or hermaphrodites.

Obstacle ability: The Mark V could cross 10-foot trenches; ford streams 3 feet deep; climb a 35-degree slope; and scale walls up to 4 feet, 11 inches high.

Crew: 8

British Mark V Star

Dimensions: Length, 32 feet, 5 inches; width, (male) 13 feet, 6 inches, (female) 10 feet, 6 inches; height, 8 feet, 8 inches
Armor plate: 0.24 to 0.59 inches thick
Weight: Male, 37 tons; female, 36 tons
Engine: Ricardo 6-cylinder, 150-horsepower, forced-water cooled
Transmission: Planetary with 4 forward and 4 reverse speeds
Horsepower/ton ratio: Male, 4.1; female, 4.2
Tracks: Flat steel plates 26½ inches wide, with single grousers
Suspension: Rigid with rollers
Fuel capacity: 111 gallons
Range: 40 miles
Maximum speed: 4 miles per hour
Armament: Male, 2 6-pounder (57mm) guns and 5 8mm machine guns; female 7 8mm machine guns
Obstacle ability: The Mark V Star could cross 14-foot trenches; ford streams 3 feet, 3 inches deep; climb a 30-degree slope; scale walls up to 4 feet, 10 inches high; and knock down a 22-inch-diameter tree.
Crew: 8

United States Mark VIII Liberty

Dimensions: Length, 34 feet, 2½ inches; width, 12 feet, 5 inches; height, 10 feet, 2½ inches
Armor plate: 0.236 to 0.63 inches thick
Weight: 43.5 tons combat loaded
Engine: Liberty 12-cylinder, V-type, 338-horsepower, forced-water cooled
Transmission: Planetary with 2 forward and 2 reverse speeds
Horsepower/ton ratio: 7.8
Tracks: Flat metal plates 26½ inches wide, with single grousers
Suspension: Rigid with rollers
Fuel capacity: 240 gallons
Range: 50 miles
Maximum speed: 6.5 miles per hour
Armament: 2 6-pounder (57mm) guns and 5 .30-caliber machine guns

Obstacle ability: The Liberty could cross trenches 16 feet wide, ford 2-foot-deep streams, climb a 40-degree slope, scale a 54-inch vertical wall, and knock down 18-inch-diameter trees.

Crew: 11

French Schneider Tank

Dimensions: Length, 19 feet, 8 inches; width, 6 feet, 7 inches; height, 7 feet, 10 inches

Armor plate: 0.2 to 0.95 inches thick, with double plates on the front, sides, and top. The double plates were positioned with a 1.5-inch space between them.

Weight: 14.9 tons

Engine: Schneider 4-cylinder, 70-horsepower, forced-water cooled

Transmission: Sliding gear with 3 forward speeds and 1 reverse

Horsepower/ton ratio: 4.7

Tracks: Solid plates 14 inches wide, with single grousers

Suspension: Vertical coil springs with jointed bogie frames

Fuel capacity: 53 gallons

Range: 25 miles

Maximum speed: 5 miles per hour

Armament: 1 short-barreled 75mm gun mounted in the front, and a single machine gun mounted on each side of the tank

Obstacle ability: The Schneider could cross trenches 5 feet, 10 inches deep; ford 2-foot-7-inch streams; scale a 2-foot-7-inch wall; climb a 30-degree slope; and knock down trees up to 16 inches in diameter.

Crew: 6

French St. Chaumond Tank

Dimensions: Length, 28 feet, 10 inches (including gun); width, 8 feet, 9 inches; height, 7 feet, 8 inches

Armor plate: 0.2 inches to 0.67 inches

Weight: 25.3 tons

Engine: Panhard 4-cylinder, 90-horsepower, water-cooled, with dynamo, two electric motors, and storage batteries

Transmission: Crochat-Colardeau, electric

Horsepower/ton ratio: 3.6

Tracks: Solid plates 19¾ inches wide, with double grousers

Suspension: Coil springs, rollers, and bogies

Fuel capacity: 66 gallons

Range: 37 miles

Maximum speed: 5 miles per hour

Armament: 1 75mm gun mounted in the front, and 4 machine guns (1 in front, 1 in each side, 1 in the rear)

Obstacle ability: The St. Chaumond could cross trenches 8 feet wide, ford a stream 31 inches deep, climb a 30-degree slope, scale a 15-inch wall, and knock down trees up to 15 inches in diameter.

Crew: 9

APPENDIX B

Tank Corps Personnel Cited for Valor[1]

Medal of Honor

Cpl. Donald M. Call
Cpl. Harold W. Roberts (Posthumous)

Distinguished Service Cross

Col. George S. Patton, Jr.
Col. Daniel D. Pullen
Maj. Sereno E. Brett
Capt. Math L. English (Posthumous, 2 awards)
Capt. Dean M. Gilfillan
Capt. Harry H. Semmes (2 awards)
Capt. Kit R. Varney (Posthumous)
Capt. Newell P. Weed
1st Lt. Herbert J. Edwards
1st Lt. Paul S. Edwards
1st Lt. Tom W. Saul
1st Lt. Carl J. Sonstelie
2d Lt. Harold J. Ash
2d Lt. David M. Bowes
2d Lt. John Gleason
2d Lt. Edwin A. McClure (2 awards)
2d Lt. A. B. Maury
2d Lt. Julian K. Morrison (2 awards)
2d Lt. William C. Rock (Posthumous)
2d Lt. John R. Webb
Sfc. Earnest E. Ely
Sgt. Raymond C. Chisholm (Posthumous)
Sgt. Harley R. Nichols

---

[1] Source: *301st Battalion History;* General Orders no. 2, GHQ–Tank Corps, AEF; General Orders no. 24, GHQ–Tank Corps, AEF.

Sgt. David Winton
Sgt. Charles C. Young
Cpl. Walter H. Blanchard
Cpl. William E. Brophy
Cpl. Troy C. Carroll
Cpl. Albert F. Neil
Cpl. Frank W. Osmond
Cpl. Albert J. Zimborski (Posthumous)
Pfc. Joseph T. Angelo
Pvt. Charles D. Merritt
Pvt. Ralph L. Voight
Pvt. Robert F. Wisher

British Distinguished Service Order

Maj. Ralph I. Sasse

British Military Cross

Capt. R. deP. Clarke
Capt. John M. Franklin
Capt. Theodore C. Naedele
1st Lt. M. Ferrell
1st Lt. Elliott P. Frost
2d Lt. Earl Dunning
2d Lt. E. Kusener
2d Lt. A. B. Maury

British Military Medal

Sgt. Leon Parker
Sgt. Carl Rosenhagen
Sgt. H. Ruhs
Sgt. Harry M. Sponable
Cpl. F. Kriegeskotte
Cpl. A. E. Pfaff
Pvt. Russell C. Davis
Pvt. Hubert S. Morrison
Pvt. Ralph L. Voight

British Distinguished Conduct Medal

Sgt. N. E. Jardel
Sgt. P. Monahan
Cpl. Robert J. McCulloch
Cpl. Albert F. Neil
Cpl. Frank W. Osmond
Pvt. Robert F. Wisher

British Meritorious Service Medal

Sgt. P. M. Hart
Sgt. Edward M. Noonan

British Disinguished Service Card

Sgt. Francis W. Connelly
Cpl. Lloyd L. Graeff
Pvt. G. Braun
Pvt. W. H. Milam

Selected Bibliography

Manuscript Collections

Carlisle Barracks, Pa. U.S. Army Military History Institute. Archives. World War I Survey Collection.

Fort Knox, Ky. Patton Museum of Cavalry and Armor. Archives. Joseph W. Viner Collection.

Washington, D.C. National Archives. Modern Military Records Division. Records Groups 120 and 165.

Washington, D.C. Library of Congress. Manuscript Division. George S. Patton, Jr., Collection

Unpublished Sources

Carlisle Barracks, Pa. U.S. Army Military History Institute. Carlson, R. E., Maj. "Memorandum on the Development of Tanks." 16 Mar. 1921.

———. Drain, James A., Maj., and Alden, Herbert W., Maj. "Report of Investigations by Majors Drain and Alden." 10 Nov. 1917.

———. Murphy, Frank T., 1st Lt. "The Moral Effect of Tanks upon the Enemy." 1919.

———. Nenninger, Timothy K. "The Development of American Armor, 1917–1940." M.A. thesis. University of Wisconsin, 1968.

———. Rockenbach, Samuel D., Brig. Gen. "Operations of the Tank Corps, A.E.F., with the 1st American Army." Dec. 1918.

———. Rockenbach, Samuel D., Brig. Gen. "Report of Chief of the Tank Corps." 13 Oct. 1919.

———. Rockenbach, Samuel D., Brig. Gen. "The Tank Corps: A Talk to the General Staff." 24 Sept. 1919.

———. Rockenbach, Samuel D., Col. "Tanks: A Lecture Delivered at the Tank School." 1923.

———. U.S. Army. "Tank Corps Combat." 1919.

———. U.S. Army. "Tank Corps Drill Regulations." 1919.

243

Published Sources — Books

A Company, 301st Tank Battalion History. Philadelphia: E. A. Wright Co., 1919.

Ambrose, Stephen E. *Eisenhower,* Vol. 1, *Soldier, General of the Army, President Elect, 1890–1952.* New York: Simon & Schuster, 1983.

American Battle Monuments Commission. *American Armies and Battlefields in Europe: A History, Guide, and Reference Book.* Washington, D.C.: U.S. Government Printing Office, 1938.

Ayers, Leonard P., Col. *The War with Germany: A Statistical Summary.* Washington, D.C.: U.S. Government Printing Office, 1919.

Blumenson, Martin. *The Patton Papers, 1885–1940.* Boston: Houghton Mifflin Co., 1972.

Braim, Paul F. *The Test of Battle: The American Expeditionary Forces in the Meuse-Argonne Campaign.* Newark, N.J.: University of Delaware Press, 1987.

Bullard, Robert L., Major General. *Personalities and Reminiscences of the War.* Garden City, N.Y.: Doubleday, Page & Co., 1925.

Coffman, Edward M. *The War to End All Wars: The American Military Experience in World War I.* New York: Oxford University Press, 1968.

Crowell, Benedict, and Robert F. Wilson. *The Armies of Industry: Our Nation's Manufacture of Munitions for a World in Arms, 1917–1918.* Vol. I. New Haven, Conn.: Yale University Press, 1921.

Edmonds, Sir James E., Brig. Gen., and Lt. Col. R. Maxwell-Hyslop, eds. *History of the Great War.* Vol. V: *Military Operations France and Belgium 1918 — 26th September to 11th November: The Advance to Victory.* London: His Majesty's Stationery Office, 1947.

Eisenhower, Dwight D. *At Ease: Stories I Tell to Friends.* Garden City, N.Y.: Doubleday & Company, Inc., 1967.

Fuller, John F. C., Major General. *Tanks in the Great War, 1914–1918.* London: John Murray, 1920.

Gillie, Mildred H. *Forging the Thunderbolt; A History of the Development of the Armored Force.* Harrisburg, Pa.: Military Service Publishing Company, 1947.

Handbook of the Six-Ton Special Tractor Model 1917. Washington, D.C.: U.S. Government Printing Office, 1918.

Jones, Ralph E., Maj.; Capt. George H. Rarey and 1st Lt. Robert J. Icks. *The Fighting Tanks Since 1916.* Washington, D.C.: The National Service Publishing Co., 1933.

Kenamore, Clair. *From Vauquois Hill to Exermont: A History of the Thirty-Fifth Division of the United States Army.* St. Louis, Mo.: Guard Publishing Company, 1919.

Liggett, Hunter, Lt. Gen. *AEF: Ten Years Ago in France.* New York: Dodd, Mead & Co., 1928.

Macksey, Kenneth, ed. *Tank Facts and Feats: A Record of Armoured Fighting Vehicle Achievement.* New York: The Two Continents Publishing Group, Ltd., 1974.

——, and John Batchelor. *Tank: A History of the Armoured Fighting Vehicle.* New York: Charles Scribner's Sons, 1970.

Maurer, Maurer, ed. *The United States Air Service in World War I: The Final Report and a Tactical History,* Vol. I. Washington, D.C.: Office of Air Force History, 1978.

Miller, Merle. *Ike the Soldier: As They Knew Him.* New York: G. P. Putnam's Sons, 1987.

Millett, Allan R., and Peter Maslowski. *For the Common Defense: A Military History of the United States of America.* New York: The Free Press, a division of Macmillan, Inc., 1984.

Mitchell, F. *Tank Warfare: The Story of Tanks in the Great War.* London: Thomas Nelson & Sons, Ltd. 1933.

Ogorkiewicz, Richard M. *Armor: A History of Mechanized Forces.* New York: Frederick A. Praeger, 1960.

Order of Battle of the United States Land Forces in the World War (1917–19), Zone of the Interior. Vol. 3 Part 2. Washington, D.C.: U.S. Government Printing Office, 1949.

O'Ryan, John F., Major General. *The Story of the 27th Division.* 2 vols. New York: Wynkoop Hallenbeck Crawford Co., 1921.

Palmer, Frederick. *Newton D. Baker: America at War,* Vol. 2. New York: Dodd, Mead & Co., 1931.

Paxson, Frederic L. *American Diplomacy and the World War: America at War,* Vol. 2. New York: Cooper Square Publishers, 1966. Reprinted by permission of Houghton Mifflin Co., Boston, copyright 1939.

Pershing, John J., General. *My Experience in the World War.* 2 vols. New York: Stokes, 1931.

Personnel Specifications, Tank Corps. Washington, D.C.: War Department, The Adjutant General's Office, Classification Division, 1918.

Preliminary Handbook of the Mark VIII Tank. Ordnance Department Pamphlet no. 1977. Washington, D.C.: U.S. Government Printing Office, 1925.

Semmes, Harry H. *Portrait of Patton.* New York: Appleton-Century-Crofts, Inc., 1955.

Smythe, Donald. *Pershing: General of the Armies.* Bloomington, Ind.: Indiana University Press, 1986.

Society of the First Division. *History of the First Division During the World War, 1917–1919.* Philadelphia: The John C. Winston Co., 1922.

Stallings, Laurence. *The Doughboys: The Story of the AEF, 1917–1918.* New York: Harper & Row Publishers, 1963.

Stubbs, Mary L., and Stanley R. Connor. *Army Lineage Series: Armor-Cavalry.* Part I. Washington, D.C.: Office of the Chief of Military History, U.S. Army, 1969.

Sueter, Murray, Rear Adm. Sir. *The Evolution of the Tank.* London: Hutchinson & Company, Ltd., 1937.

Tanks: Organization and Tactics. Washington, D.C.: War Department, AEF Pamphlet no. 1432 G-5, Dec., 1918.

Thomas, Shipley, Capt. *The History of the AEF.* New York: George H. Doran Co., 1920.

U.S. Army in the World War, 1917–1918: Organization of the AEF, Vol. 1. Washington, D.C.: Historical Division, U.S. Army, 1948.

Vandiver, Frank E. *Black Jack: The Life and Times of John J. Pershing,* Vol. 2. College Station, Tex.: Texas A&M University Press, 1977.

Van Every, Dale. *The A.E.F. in Battle.* New York: D. Appleton & Company, 1928.

Viner, Joseph W., Capt. *Tactics and Techniques of Tanks.* Fort Leavenworth, Kan.: The General Service Schools Press, 1920.

Weigley, Russell F. *History of the United States Army, Enlarged Edition.* Bloomington, Ind.: Indiana University Press, 1984.

Wilson, G. Murray, ed. *Fighting Tanks.* London: Seeley, Service and Co., 1929.

Whitehouse, Arch. *Tank: The Story of Their Battles and the Men Who Drove Them from Their First Use in WWI to Korea.* Garden City, N.Y.: Doubleday & Co., Inc., 1960.

Who Was Who in American History—The Military. Chicago: Marquis Who's Who, Inc., 1975.

Published Sources—Articles

Chynoweth, Bradford G., Maj. "Cavalry Tanks." *Cavalry Journal* 30 (July 1921): 247-51.

Eisenhower, Dwight D., Capt. "A Tank Discussion," *Infantry Journal* 17, no. 5 (Nov. 1920): 453-58.

Kraft, William R., Jr., Maj. Gen. (U.S. Army, Ret.). "The Saga of the Five of Hearts." *Armor* 97 (Jul.-Aug. 1988): 35-38.

Nenninger, Timothy K. "The Tank Corps Reorganized." *Armor* 78, no. 2 (Mar.-Apr. 1969): 34-38.

——. "The World War I Experience." *Armor* 78, no. 1 (Jan.-Feb. 1969): 46-51.

Patton, George S., Jr., Col. "Comments on 'Cavalry Tanks.'" *Cavalry Journal* 30 (July 1921): 251-52.

——. "Tanks in Future Wars." *Infantry Journal* 16, no. 11 (May 1920): 958-62.

Rockenbach, Samuel D., Brig. Gen. "Cooperation of Tanks with Other Arms." *Infantry Journal* 16, nos. 7 and 8 (Jan. and Feb. 1920): 533-45, 662-73.

Rogge, Robert E. "The 304th Tank Brigade: Its Formation and First Two Actions." *Armor* 97 (Jul.-Aug. 1988): 26-34.

Rosenhagen, Carl. "A Day in the Life of a Tanker." *Armor* 82, no. 5 (Sept.-Oct. 1973): 34-35.

Schreier, Konrad F., Jr. "The American Six-Ton Tank." *Armor* 77 (Nov.-Dec. 1968): 45-49.

Index